Mathematics for Engineers

Ordinary differential equations

W.Bolton

Longman Scientific & Technical,
Longman Group UK Limited,
Longman House, Burnt Mill, Harlow,
Essex, CM20 2JE, England
and Associated Companies throughout the world.

First published 1994

ISBN 0 582 233992

British Library Cataloguing in Publication Data
A CIP record for this book is available from the British Library

Printed in Malaysia by TCP

Contents

Preface

This is one of the books in a series designed to provide engineering students in colleges and universities with a mathematical toolkit. In the United Kingdom it is aimed primarily at HNC/HND students and first-year undergraduates. Thus the mathematics assumed is that in BTEC National Certificates and Diplomas or in A level. The pace of development of the mathematics has been aimed at the notional reader for whom mathematics is not their prime interest or "best subject" but need the mathematics in their other studies. The mathematics is developed and applied in an engineering context with large numbers of worked examples and problems, all with answers, being supplied.

This book is concerned with ordinary differential equations and also their application in, primarily, electrical/electronic and mechanical engineering. A familiarity with basic calculus is assumed. The aim has been to include sufficient worked examples and problems to enable the reader to acquire some proficiency in the handling of ordinary differential equations and their use in engineering.

W.Bolton

1 Introducing differential equations

1.1 Introduction

Differential equations occur with numerous problems encountered in engineering, in science and in many other spheres of life. For example, they occur in such diverse problems as

1 the motion of projectiles,
2 the conduction of heat in a solid,
3 the cooling of a solid or liquid,
4 oscillations with mechanical or electrical systems,
5 the deflections of a beam,
6 the transient currents and voltages in electrical circuits,
7 the rate of decay of radioactive materials,
8 the growth of populations, whether people or bacteria,
9 chemical reactions.

This chapter is a general look at such equations in engineering and the terminology associated with them. Later chapters in this book consider differential equations, and their applications, in more detail.

1.2 Rates of change

Consider the relation between force and acceleration given by Newton's second law, i.e.

force = mass × acceleration

Acceleration is the rate of change of velocity with time and thus we can write

force = mass × rate of change of velocity

We have a relationship between a quantity and the rate of change of some other quantity. Such a type of relationship is quite

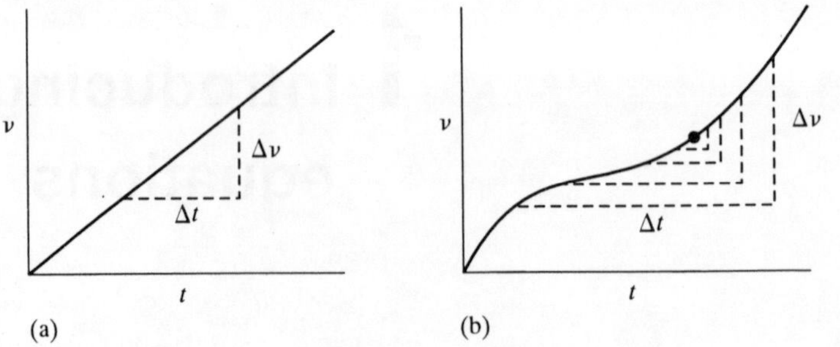

Fig. 1.1 Rate of change of velocity with time

(a) (b)

common in engineering, and many other subjects.

Consider a graph of how the velocity varies with time when there is a constant force, as in figure 1.1(a). A constant force requires a constant rate of change of velocity, assuming the mass remains constant, and thus the slope of the velocity–time graph is constant. Then the rate of change of velocity with time is

$$\text{rate of change of velocity} = \frac{\Delta v}{\Delta t}$$

where Δv is the change in the velocity occurring in a time Δt, the Δ symbol placed in front of the velocity or time symbols being used to indicate that we are concerned with an increment of velocity or time. Thus if the velocity changes from 2 m/s to 8 m/s in 3 s then the rate of change of velocity is

$$\text{rate of change of velocity} = \frac{8-2}{3} = 2 \text{ m/s}^2$$

Where the force is not constant, as in figure 1.1(b) with the slope of the graph changing, then $\Delta v/\Delta t$ gives the average rate of change of velocity with time over the time interval concerned. Thus if the velocity changes from 2 m/s to 8 m/s in 3 s then, when the force is not constant, we can only say that the average rate of change of velocity over the 3 s is 2 m/s^2, we cannot state what the rate of change of velocity with time will be at some particular time during that time interval. A more reasonable approximation to that would be obtained if we knew how the velocity changed in going from 0.1 s before the time to 0.1 s after it. The smaller we make the time interval over which we consider the velocity change the closer it comes to giving the instantaneous rate of change of velocity with time at the time concerned.

$\Delta v/\Delta t$ becomes equal to the slope of the graph at a point when we make the time increment Δt vanishingly small. When we do this we write $\Delta v/\Delta t$ as dv/dt. This is termed the *derivative*.

$$\frac{dv}{dt} = \lim_{\Delta t \to 0} \frac{\Delta v}{\Delta t} \qquad [1]$$

Thus the rate of change of velocity v with time t, at some instant of time, is represented by

$$\text{rate of change of velocity with time} = \frac{\mathrm{d}v}{\mathrm{d}t}$$

The process by which the gradient function is obtained is known as *differentiation*. The following example illustrates the above discussion of obtaining the gradient by working with finite time intervals and then considering the limit as the time interval is reduced to zero. You can check the result by the use of the standard rules of differential calculus. The following are some commonly used standard forms:

$$\frac{\mathrm{d}}{\mathrm{d}x}(mx^n) = nmx^{n-1} \qquad [2]$$

$$\frac{\mathrm{d}}{\mathrm{d}x}(\sin ax) = a\cos ax \qquad [3]$$

$$\frac{\mathrm{d}}{\mathrm{d}x}(\cos ax) = -a\sin ax \qquad [4]$$

$$\frac{\mathrm{d}}{\mathrm{d}x}(\mathrm{e}^{ax}) = a\,\mathrm{e}^{ax} \qquad [5]$$

Example

Consider an object falling freely from rest with a constant acceleration of g. The distance x fallen from rest is given by $x = \frac{1}{2}gt^2$. Derive an equation for the instantaneous velocity v after a time t.

Consider the body to have fallen a distance x after a time t and a distance $x + \Delta x$ after a time $t + \Delta t$. Then

$$x + \Delta x = \tfrac{1}{2}g(t + \Delta t)^2$$

$$= \tfrac{1}{2}[t^2 + 2t\,\Delta t + (\Delta t)^2]$$

Since $x = \frac{1}{2}gt^2$, then

$$\Delta x = gt\,\Delta t + \tfrac{1}{2}g(\Delta t)^2$$

The average velocity over the time interval Δt is thus

$$\text{average velocity} = \frac{\Delta x}{\Delta t} = gt + \tfrac{1}{2}g\,\Delta t$$

The instantaneous velocity after a time t is thus

$$v = \frac{dx}{dt} = \lim_{\Delta t \to 0} (gt + \tfrac{1}{2}g\,\Delta t) = gt$$

The same result could have been obtained by differentiating $x = \frac{1}{2}gt^2$, i.e. using equation [2].

Review problems

1 The way in which the displacement x of a particle varies with time t is given by the following equations. Derive an equation for the instantaneous velocity v, i.e. dx/dt, at a time t in each case.

(a) $x = 2t^2 + 3t - 4$, (b) $x = 3 \sin 2t$, (c) $x = 5\,e^{-2t}$

2 A moving iron ammeter has a non-linear scale. The deflection θ for a current I is given by the relationship

$\theta = cI^2$

Derive an equation for the current sensitivity $d\theta/dI$ at a particular current I.

1.2.1 Rate of change of rate of change

Since velocity is the rate of change of displacement x with time t, we could write for the velocity v at some instant of time

$$v = \frac{dx}{dt}$$

But acceleration a is the rate of change of velocity with time, i.e.

$$a = \frac{dv}{dt}$$

Thus, acceleration is the rate of change of the rate of change of displacement with time, i.e.

$$a = \frac{d}{dt}\left(\frac{dx}{dt}\right)$$

This is written as

$$a = \frac{d^2x}{dt^2}$$

Note that d^2x/dt^2 is not dx/dt squared. One is the rate of change of a rate of change while the other is the square of a rate of change.

Example

The displacement x of an object with time t is given by the equation

$$x = 4t^2 + 2t - 3$$

Hence derive an equation for d^2x/dt^2, i.e. the acceleration.

Differentiating the above equation gives

$$\frac{dx}{dt} = 8t + 2$$

Differentiating again gives

$$\frac{d^2x}{dt^2} = 8$$

Review problems

3 Derive the values of dx/dt and d^2x/dt^2 for the displacement x when it is related to time t by:

(a) $x = 5t^2 + 2t$, (b) $x = 2t^3 + 3t^2 + 4t + 5$,

(c) $x = 4 \sin 3t$, (d) $x = 2\,e^{-3t}$

1.2.2 Algebraic sign of derivatives

If we have y as some function of x then dy/dx is positive if a small increase in the value of x produces a small increase in the value of y. On a graph of y against x then this means that the slope of the graph is positive, as in figure 1.2(a). A graph of y against x which has a negative slope and hence negative values of dy/dx is shown in figure 1.2(b).

d^2y/dx^2 is positive if a small increase in x produces a small increase in the value of dy/dx. On a graph of y against x then this means that as x increases then the slope of the graph should

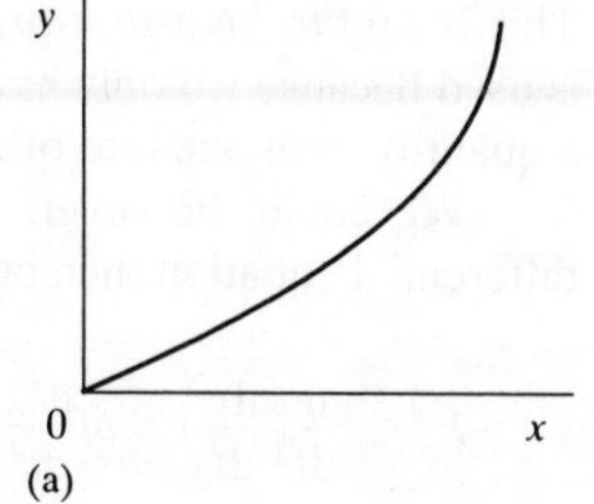

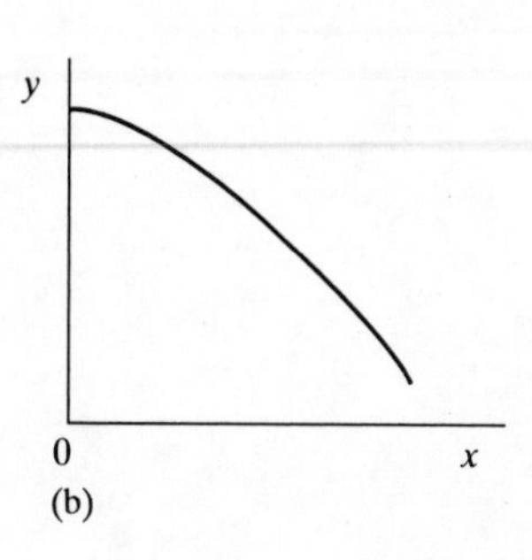

Fig. 1.2 (a) Positive values of dy/dx, (b) negative values of dy/dx

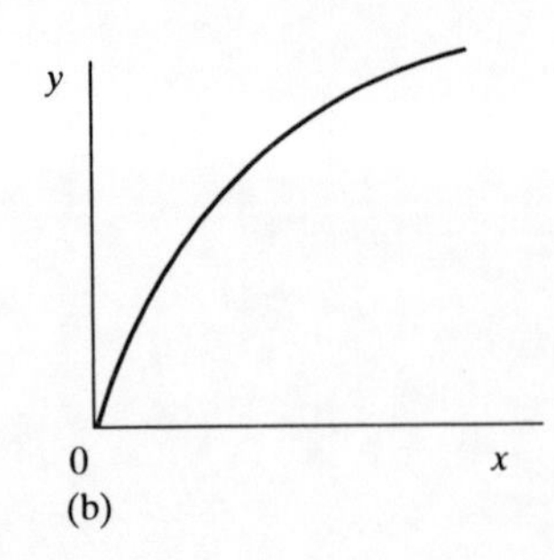

Fig. 1.3 (a) Positive values of d^2y/dx^2, (b) negative values of d^2y/dx^2

increase if d^2y/dx^2 is to be positive, as in figure 1.3(a). In figure 1.3(b) the slope of the graph decreases as x increases and so d^2y/dt^2 is negative.

Example

In radioactive decay the number N of radioactive atoms decreases with time t. The activity is the rate at which the number of radioactive atoms changes, i.e. $A = dN/dt$. Is dN/dt positive or negative?

A small increase in time produces a small decrease in N. Hence dN/dt is negative.

Review problems

4 When a projectile is thrown upwards, the velocity v decreases as the height h above the starting point increases. Heights measured in an upwards direction are considered to be positive. The velocity is the rate of change of height with time. Is dh/dt positive or negative?

1.3 Differential equations

An equation involving derivatives is called a *differential equation*. Thus, for example, Newton's second law can be written in the form of a differential equation as

$$F = m\frac{dv}{dt}$$

This is said to be a *first-order* differential equation. The term first is used because we only have in the equation a derivative involving a quantity with one rate of change.

We could, however, write the velocity as dx/dt and so the differential equation then becomes

$$F = m\frac{d}{dt}\left(\frac{dx}{dt}\right) = m\frac{d^2x}{dt^2} \qquad [6]$$

This is termed a *second-order differential equation* since the highest derivative is a rate of change of a rate of change.

With an equation such as $y = 3x$, then y is a function of x. We can represent this by writing y as $y = f(x)$. Since y and x can have a number of possible values, they are called *variables*. y is termed the *dependent variable* and x the *independent variable*. This means that any value can be given to x and the corresponding value of y will be determined from it. With a differential equation, the variable being differentiated is the dependent variable. Thus we have, in this case, dy/dx.

Equations containing integrals are called *integral equations* and those with both derivatives and integrals are *integro-differential equations*.

Example

Write the differential equation which can represent the following situation: the rate at which the temperature of a liquid cools is proportional to the temperature difference between that of the liquid θ and its surroundings θ_s.

The rate of cooling of the liquid can be represented by $d\theta/dt$, and so

$$\frac{d\theta}{dt} = k(\theta - \theta_s)$$

where k is the constant of proportionality. The above relationship is known as *Newton's law of cooling*.

Example

Simple harmonic motion is an oscillation where the acceleration of the oscillating body is always proportional to its displacement from its rest position and directed towards it. Write a differential equation to describe this motion.

Acceleration is the rate of change of velocity with time, with velocity being the rate of change of displacement with time. Thus acceleration is

$$\text{acceleration} = \frac{dv}{dt} = \frac{d}{dt}\left(\frac{dx}{dt}\right) = \frac{d^2x}{dt^2}$$

With the acceleration being proportional to the displacement x we have the acceleration equal to kx, where k is the constant of proportionality. Because the acceleration is always directed towards the rest position, we have a situation where the velocity of the object decreases as x increases, i.e. as an object moves out from the rest position the acceleration is a retardation. This means a negative value for the acceleration (see section 1.2.2). Thus the differential equation is

$$\frac{d^2x}{dt^2} = -kx$$

Review problems

5 Write differential equations to represent each of the following situations:
(a) The potential difference v across an inductor is proportional to the rate of change of current i through it, the constant of proportionality being the inductance L.
(b) The damping force F experienced by an object is proportional to the rate at which the displacement x of the object changes with time.
(c) The current i is equal to the rate of movement of charge q.
(d) The rate at which the angular velocity of a body changes with time is proportional to the torque T acting on it, the constant of proportionality being the moment of inertia I.
(e) At a constant temperature, the rate of change of the volume V of an ideal gas with respect to the pressure p is proportional to $-V/p$ (note that this is just the differentiation of Boyle's law).
(f) The thickness x of ice on a frozen lake increases at a rate which is proportional to the square root of the temperature of the air θ in °C.

1.3.1 Modelling with differential equations

We can talk of the differential equations, or indeed any equations, which are needed to specify the behaviour of a system as being a *mathematical model* of the system. Mathematical modelling can be considered to involve a number of stages:

1 The real world situation is considered and formulated in mathematical terms, i.e. a mathematical model is constructed.
2 Mathematical analysis can be carried out on the model.
3 The results of the analysis can then be used to lead to an interpretation in the context of the real world situation.

The following examples illustrate how such models may be constructed.

Example

Derive a mathematical model for an electrical circuit consisting of

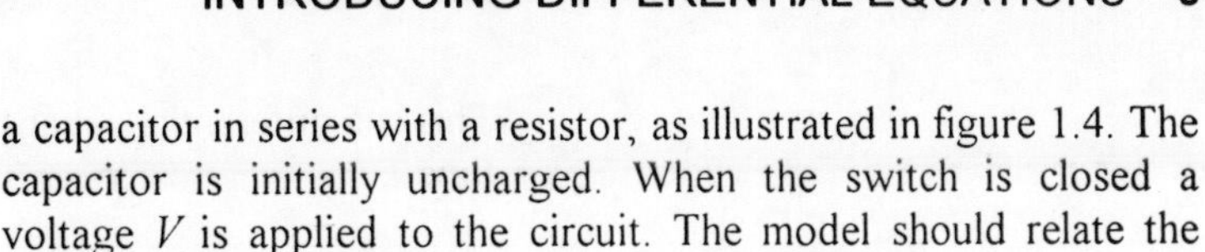

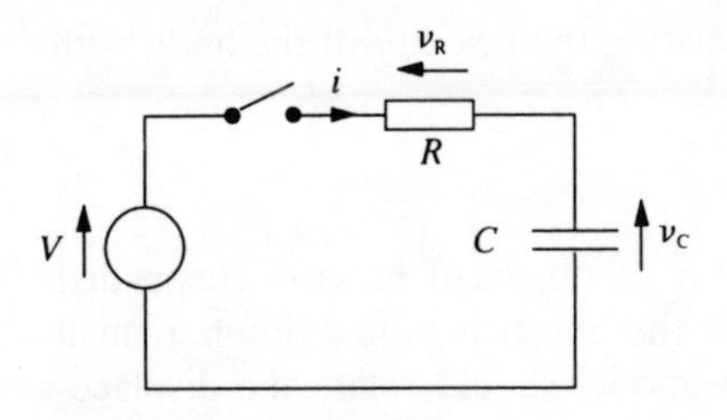

Fig. 1.4 *RC* circuit

a capacitor in series with a resistor, as illustrated in figure 1.4. The capacitor is initially uncharged. When the switch is closed a voltage V is applied to the circuit. The model should relate the potential difference across the capacitor with time.

At any instant we can, using Kirchhoff's voltage law, write for the potential differences

$$v_R + v_C = V$$

where v_C is the potential difference across the capacitor and v_R that across the resistor. But $v_R = Ri$ and the current i in the circuit is $C(\mathrm{d}v_C/\mathrm{d}t)$, hence the above equation can be written as

$$RC\frac{\mathrm{d}v_C}{\mathrm{d}t} + v_C = V$$

We thus have a differential equation relating the potential difference across the capacitor with time. The equation is first order.

Example

Derive a mathematical model for a body of mass m freely falling in air (figure 1.5) and which relates its velocity with time.

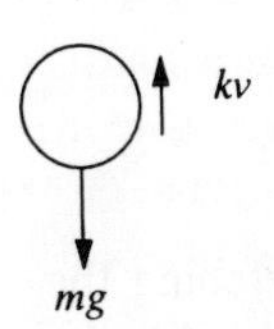

Fig. 1.5 Body falling in air

The gravitational force acting downwards on the body is mg, where g is the acceleration due to gravity. Opposing the movement of the body through the air is a force due to air resistance. We will assume that this air resistance force is proportional to the velocity v of the body, i.e. is kv where k is a constant. Thus the net force F acting on the body is

$$F = mg - kv$$

But Newton's second law gives

$$F = ma$$

where a is the acceleration of the body, i.e. its rate of change of velocity. Thus

$$m\frac{\mathrm{d}v}{\mathrm{d}t} = mg - kv$$

and so we can write

$$m\frac{\mathrm{d}v}{\mathrm{d}t} + kv = mg$$

This is a differential equation relating the velocity of the body with time.

Example

Derive a mathematical model for an object of mass m suspended from a spring (figure 1.6) when the object is pulled down a small amount and then released. The model should relate the displacement of the object with time.

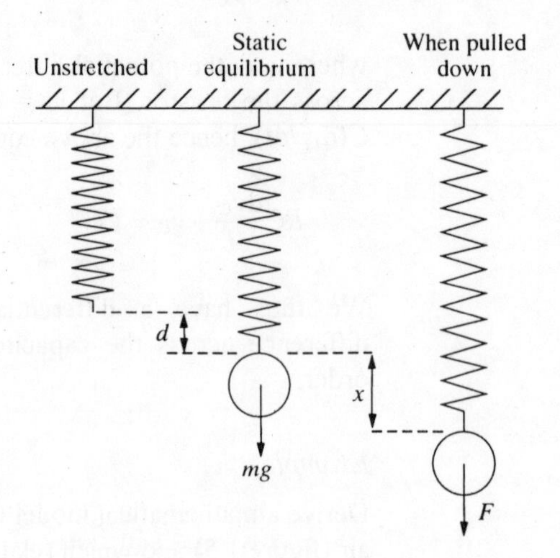

Fig. 1.6 Mass on spring

The gravitational force acting on the mass, and so stretching the spring, is mg. The mass is then in static equilibrium under the action of this force mg and the upwards force acting on the body due to the stretching of the spring. If the spring extends by an amount d then, since the force exerted by the spring is proportional to its extension (Hooke's law), we have

$$mg = kd$$

where k is a constant. Now we pull the mass down by an amount x from this static equilibrium position. When the mass is released and allowed to move, then the total force acting on it is

$$F = mg - kd - kx$$

Since $mg = kd$, then

$$F = -kx$$

But, according to Newton's second law $F = ma$, where a is the acceleration of the mass. But acceleration is rate of change of velocity and velocity is rate of change of displacement. Thus

$$m\frac{d^2x}{dt^2} = -kx$$

This can be written as

$$m\frac{d^2x}{dt^2} + kx = 0$$

This is a differential equation relating the displacement of the mass with time and describes the free oscillations of the mass.

Example

Derive a mathematical model for a situation involving liquid entering and leaving from an open tank, as in figure 1.7, and which relates the height of liquid with time.

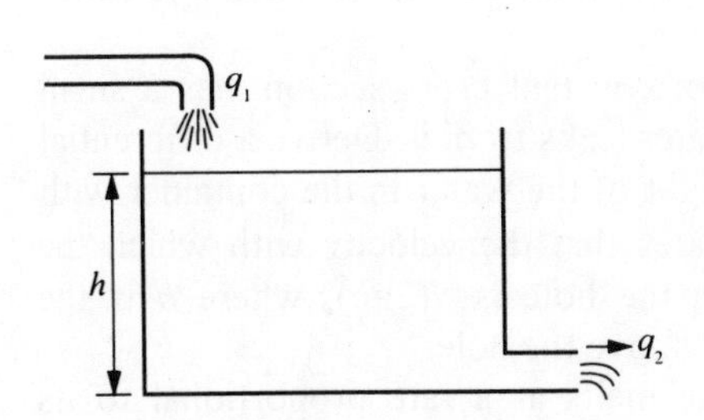

Fig. 1.7 Liquid in a container

If the liquid enters the tank at the volumetric rate of q_1 and leaves it at the rate of q_2, then the rate at which the volume V of liquid in the container changes is

$$\frac{dV}{dt} = q_1 - q_2$$

But $V = Ah$, where A is the uniform cross-sectional area of the container and h the height of the liquid in the container. Thus

$$\frac{d(Ah)}{dt} = A\frac{dh}{dt} = q_1 - q_2$$

The rate q_2 at which the liquid leaves the container, when flowing out into the atmosphere, is given by *Torricelli's theorem* (or deduced from Bernoulli's equation)

$$q_2 = \sqrt{2gh}$$

where g is the acceleration due to gravity. Thus

$$A\frac{dh}{dt} + \sqrt{2gh} = q_1$$

This is a differential equation describing how the height of the liquid in the container will vary with time.

Review problems

6 Write a differential equation for each of the processes outlined below:

(a) An object of mass m is thrown vertically upwards. The air

resistance is proportional to the square of the velocity. Derive the differential equation relating the velocity v and time.
(b) A skydiver falls from a plane towards the earth. Before the parachute opens the air resistance is proportional to the velocity v of fall. If the skydiver plus equipment has a total mass of m, derive the differential equation relating the velocity and time of fall.
(c) An electrical circuit consists of a resistance R in series with an inductance L. When a current i flows through an inductor then the potential difference across the inductor is $L(\mathrm{d}i/\mathrm{d}t)$. Hence derive the differential equation relating the current in the circuit and time when a voltage of V is switched into series with the two components.
(d) A water container, of constant cross-section, has a small hole near its base and water leaks from it. Derive a differential equation relating the height of the water in the container with time. Torricelli's law states that the velocity with which the water will emerge from the hole is $\sqrt{(2gh)}$, where h is the height of the water level above the hole.
(e) A spherical hailstone melts at a rate proportional to its surface area. Derive a differential equation relating its volume V and time.
(f) A mass m is attached to the lower end of a vertical spring. The mass is then pulled down and released. Derive a differential equation relating the displacement x of the spring, from its static equilibrium position with the weight attached, and time if the movement of the mass is subject to a damping force proportional to its velocity.

1.4 Solving differential equations

Consider the differential equation

$$\frac{\mathrm{d}y}{\mathrm{d}x} = 2$$

It describes a straight line graph with a gradient of 2. But we can draw many such lines, figure 1.8 showing some of them. All the family of lines have an equation of the form

$$y = 2x + k$$

where k is a different constant for each of the lines. The term *solution* of a differential equation is used for the relationship between the dependent and independent variables such that the differential equation is satisfied for all values of the independent variable. Thus the above differential equation has many solutions, the equation $y = 2x + k$ being referred to as the *general solution*.

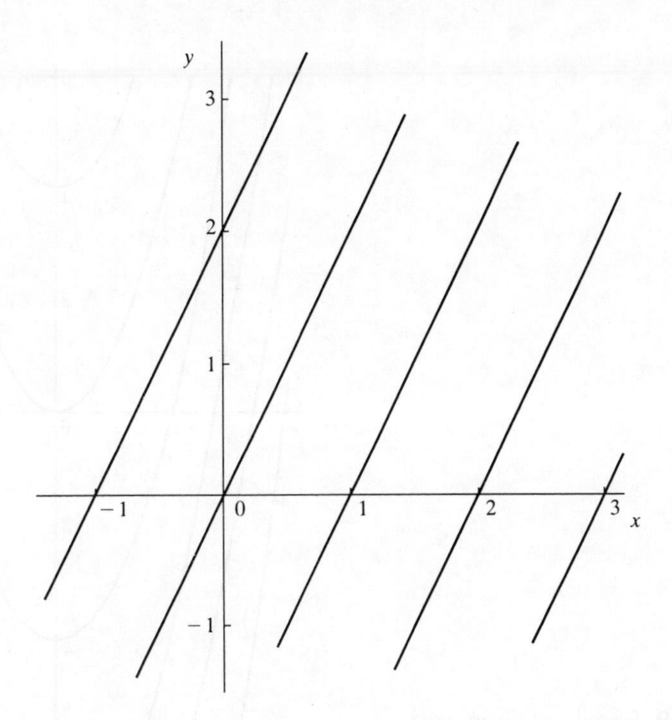

Fig. 1.8 Family of lines with d*y*/d*x* = 2

Only if adequate constraints are specified will there be only one solution, i.e. a *particular solution*. Thus if, in this case, we specify $y = 0$ when $x = 0$ then the only solution which will satisfy this condition is $y = 2x$. The term *boundary conditions* is often used for the constraints.

As another example, consider the differential equation

$$\frac{dy}{dx} = 2x$$

The general solution of this equation is

$$y = x^2 + k$$

Figure 1.9 shows some of the family of lines on a graph. If we have the condition that when $y = -3$ we have $x = 1$, then the particular solution is

$$y = x^2 - 4$$

This is one of the lines shown on the graph in figure 1.9.

Example

Verify that $y = 2 \sin 3x$ is a solution of the differential equation

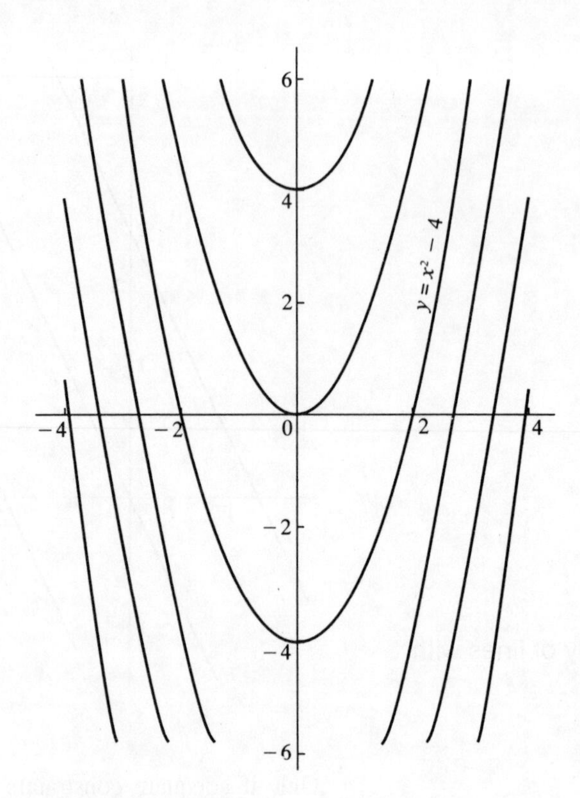

Fig. 1.9 Family of lines with $dy/dx = 2x$

$$\frac{d^2y}{dx^2} + 9y = 0$$

Differentiating y with respect to x, equation [2], gives

$$\frac{dy}{dx} = 6\cos 3x$$

Differentiating again (equation [3]) gives

$$\frac{d^2y}{dx^2} = -18\sin 3x$$

Hence, substituting in the differential equation gives

$$-18\sin 3x + 9 \times 2\sin 3x = 0$$

Hence the solution is verified.

Example

Show that $y = 6\,e^{2x} + x^3$ is a solution of the differential equation

$$\frac{dy}{dx} = 12\,e^{2x} + 3x^2$$

Differentiating y with respect to x (equations [2] and [5]) gives

$$\frac{dy}{dx} = 6 \times 2\,e^{2x} + 3x^2$$

and so confirms the solution.

Example

Given that $y = A\,e^x + B\,e^{2x}$ is a general solution of the differential equation

$$\frac{d^2y}{dx^2} - 3\frac{dy}{dt} + 2y = 0$$

find the particular solution for which $y = 3$ when $x = 0$ and $dy/dx = 5$ when $x = 0$.

Applying the first condition gives

$$3 = A\,e^0 + B\,e^0 = A + B \qquad [7]$$

Since, for the general solution when differentiated, we have

$$\frac{dy}{dx} = A\,e^x + 2B\,e^{2x}$$

then applying the second condition gives

$$5 = A\,e^0 + 2B\,e^0 = A + 2B \qquad [8]$$

Equation [7] gives $A = 3 - B$ and substituting this into equation [8] gives

$$5 = 3 - B + 2B$$

and so $B = 2$ and consequently $A = 1$. Thus the particular solution is

$$y = e^x + 2\,e^{2x}$$

Example

The differential equation describing the current in a circuit containing inductance in series with resistance, when a voltage V is applied, is

$$L\frac{di}{dt} + Ri = V$$

and has the general solution

$$-\frac{L}{R}\ln(V - Ri) = t + A$$

where A is a constant. If $i = 0$ at $t = 0$, determine the particular solution.

Substituting the constraint into the general solution equation gives

$$-\frac{L}{R}\ln V = A$$

and so

$$-\frac{L}{R}\ln(V - Ri) = t - \frac{L}{R}\ln V$$

$$\frac{L}{R}\ln\frac{V}{V - Ri} = t$$

$$\frac{V}{V - Ri} = e^{Rt/L}$$

$$i = \frac{V}{R}(1 - e^{-Rt/L})$$

Review problems

7 Verify that the functions given are solutions of the given differential equations.

(a) $y = \sin 2x$ for $\dfrac{d^2y}{dx^2} + 4y = 0$,

(b) $y = x^3 + 5$ for $\dfrac{dy}{dx} = 3x^2$,

(c) $y = e^{3x}$ for $\dfrac{d^2y}{dx^2} - 9y = 0$

8 For each of the following, verify that the solution given satisfies the differential equation and determine the value of the constants.

(a) $y = A\,\mathrm{e}^{-x}$ for $\dfrac{\mathrm{d}^2y}{\mathrm{d}x^2} + y = 0$

given $y = 2$ when $x = 0$,

(b) $y = A\sin\omega t + B\cos\omega t$ for $\dfrac{\mathrm{d}^2y}{\mathrm{d}x^2} + \omega^2 y = 0$

given $y = 4$ when $t = 0$ and $\dfrac{\mathrm{d}y}{\mathrm{d}t} = 3$ when $t = 0$,

(c) $y = 2t^3 + A$ for $\dfrac{\mathrm{d}^2y}{\mathrm{d}t^2} = 18t$

given $y = 2$ when $t = 0$,

(d) $y = t^2 + A$ for $\dfrac{\mathrm{d}y}{\mathrm{d}t} = 2t$

given $y = 5$ when $t = 0$

1.5 Notation

Derivatives are often denoted by primes. Thus we have $\mathrm{d}y/\mathrm{d}x$ denoted by y' and $\mathrm{d}^2y/\mathrm{d}x^2$ by y''. Thus for example

$$3\frac{\mathrm{d}^2y}{\mathrm{d}x^2} + 2\frac{\mathrm{d}y}{\mathrm{d}x} + 5y = 2$$

becomes

$$3y'' + 2y' + 5y = 2$$

When we have $\mathrm{d}y/\mathrm{d}t$ then the system sometimes used is to place dots above the variable. Thus $\mathrm{d}y/\mathrm{d}t$ becomes $\dot{y}$ and $\mathrm{d}^2y/\mathrm{d}t^2$ becomes $\ddot{y}$.

Review problems

9 Write out in full the following differential equations:

(a) $y'' + 2y = 0$, (b) $3y'' + 2y' + 4y = 0$

1.6 Terminology

This book is about *ordinary differential equations*. The term ordinary is used because the differential equations only contain derivatives involving the differentiation of a function with respect to a single variable. For example, the volume of a sphere is a function of the radius r of the sphere. Thus the rate of change of the volume of the sphere with time will only include $\mathrm{d}r/\mathrm{d}t$, the radius being the only function which is changing with time. The term *partial differential equation* is used when the equation contains derivatives involving the differentiation of a function with respect to two or more variables. For example, the volume of a cone is a function of both the base radius r and the cone height h. Thus if both the radius and the height are changing with time, the differential equation describing how the cone volume changes with time will involve both $\mathrm{d}r/\mathrm{d}t$ and $\mathrm{d}h/\mathrm{d}t$.

1.6.1 Classification of ordinary differential equations

Differential equations can be classified into certain groups, with each equation in a particular group solvable in the same way or ways. The following are the terms involved in such classifications.

The *order* of a differential equation is the order of the highest derivative that occurs in the equation. Thus a first-order equation contains $\mathrm{d}y/\mathrm{d}x$ and possibly y and x, while a second-order equation contains $\mathrm{d}^2y/\mathrm{d}x^2$ and possibly $\mathrm{d}y/\mathrm{d}x$, y and x. A third-order equation must contain $\mathrm{d}^3y/\mathrm{d}x^3$ as its highest derivative, a fourth-order equation $\mathrm{d}^4y/\mathrm{d}x^4$.

The *degree* of a differential equation is the power to which the highest derivative is raised. Thus, for example,

$$\left(\frac{\mathrm{d}y}{\mathrm{d}t}\right)^2 = 3y$$

has the degree of 2. It is important to realise that

$$\left(\frac{\mathrm{d}y}{\mathrm{d}t}\right)^2 \neq \frac{\mathrm{d}^2y}{\mathrm{d}t^2}$$

One is just the rate of change of y with t squared while the other is the rate of change of the rate of change of y with t.

The term *linear* is often used to describe equations and some electrical components. Thus, for example, a particular resistor may be said to be a linear device. This is because if the current through the resistor is doubled then the potential difference across it doubles, if the current is trebled then the potential difference is trebled. The equation describing the relationship between potential difference and current is $V = RI$. Another way of describing this linear property of the resistor is that if a current I_1 gives a potential difference V_1 and a current I_2 gives a potential difference V_2 then a current equal to $(I_1 + I_2)$ will give a potential difference equal to $(V_1 + V_2)$. This is often termed the *principle of superposition*. It can be used in the analysis of electrical circuits which include more than one input source. Then all but one of the sources is put to zero and, say, the current determined for that source. Then the current is determined when that source is put to zero and another source is not. The total current in the circuit with all the sources is just the sum of the currents due to each source alone. Such an analysis can only, however, be applied to circuits containing linear electrical components. Examples of non-linear electrical elements, for which the above relationships between current and potential difference cannot be applied, are thermistors, semiconductor diodes, and voltage-dependent capacitors.

If we have a linear equation, such as $y = mx$, then we can say that if $y = u + v$ then we have

$$u + v = mx$$

and we could split this into two equations,

$$u = mx \text{ and } v = 0$$

both of which present one version of the equation which must be valid. This is a technique which will be used in later chapters with differential equations.

A differential equation is said to be *linear* if

1. The dependent variable and all its derivatives occur only to the first power.
2. There are no products of terms involving the dependent variable, e.g. $y\,(\mathrm{d}y/\mathrm{d}t)$.
3. There are no functions of the dependent variable or its derivatives which are non-linear, e.g. in the form of sine or cosine or other trigonometric functions.

For example,

$$\frac{\mathrm{d}y}{\mathrm{d}t} + y = t^2$$

is linear. It does not matter that the independent variable t is raised to the power 2. The following is an example of a non-linear equation:

$$\frac{\mathrm{d}^2y}{\mathrm{d}t^2} + y^2 = 0$$

It is non-linear because of the y^2.

Linear equations can be further classified as *homogeneous* or *non-homogeneous*. When all the terms in a linear equation containing the dependent variable occur on the left-hand side of the equals sign and the terms containing the independent variable on the right-hand side, then the equation is said to be homogeneous when the terms on the right-hand side are zero and non-homogeneous when it is not. Thus, for example,

$$\frac{dy}{dt} + 5y = 0$$

is homogeneous while

$$\frac{dy}{dt} + 5y = 2t$$

is non-homogeneous, because of the $2t$.

Linear differential equations can also be classified according to the nature of the *coefficients* of the dependent variables and whether they are constant or variable. For example,

$$\frac{d^2y}{dt^2} + 3\frac{dy}{dt} + 6y = 0$$

is linear with constant coefficients, these being 1, 3 and 6 for the various terms. The following is, however, a linear equation with a variable coefficient of t:

$$\frac{d^2y}{dt^2} + t\frac{dy}{dt} + 6y = 0$$

Example

What are the orders and degrees of the following differential equations?

$$\text{(a)}\ \frac{d^2y}{dt^2} + \frac{dy}{dt} = 2t^2,\ \text{(b)}\ \left(\frac{dy}{dt}\right)^3 + y^2 = 5.$$

(a) This equation is second order because of the d^2y/dt^2 term with a degree of 1 because the highest derivative term is only to the power 1.
(b) This equation is of order 1 because of the dy/dt term with a degree of 3 because it is to the power of 3.

Example

Which of the following equations are linear?

$$\text{(a)}\ \frac{dy}{dt} + 3y = 2,\ \text{(b)}\ \frac{dy}{dt} + 2y^2 = 0,$$

(c) $\frac{dy}{dt} + \sin y = 0$, (d) $\frac{dy}{dt} + 3y = 2 \sin t$

(a) This equation is linear.
(b) This equation in non-linear, because of the y^2 term.
(c) This equation in non-linear, because of the $\sin y$ term.
(d) This equation is linear. The $\sin t$ term is not the sine of the dependent variable.

Review problems

10 What are the orders and degrees of the following differential equations?

(a) $\frac{dy}{dt} + 2y = 2t^2$, (b) $3\frac{d^2y}{dt^2} + 5\frac{dy}{dt} + 3y = t$,

(c) $\left(\frac{d^2y}{dt^2}\right)^2 = 5t$, (d) $\frac{d^2y}{dt^2} + y^2 = 0$

11 Which of the following equations are linear?

(a) $\frac{dy}{dt} + y = 2t^2$, (b) $\frac{d^2y}{dt^2} + y^2 = 4t$,

(c) $\left(\frac{dy}{dt}\right)^2 + 3y = 0$, (d) $\frac{d^2y}{dt^2} + 2\frac{dy}{dt} + 3y = t$

Further problems

12 If the displacement x of an object varies with time t according to the following equation, what will be (a) the velocity dx/dt and (b) the acceleration d^2x/dt^2?

$x = 5t^2 + 3t - 10$

13 The charge q moved through a circuit varies with time t according to the following equation. How will the current, i.e. dq/dt, vary with time?

$q = 10\,e^{-2t}$

14 The current in an electrical circuit varies with time according to the following equation. How will the rate of change of current with time vary with time?

$i = 2 \sin 5t$

15 Write differential equations to represent each of the following situations:
(a) The current i through a capacitor is proportional to the rate of change of voltage v across it, the constant of proportionality being the capacitance C.
(b) The rate at which the pressure p at the base of a container of liquid changes is proportional to the difference between the rate at which liquid enters the container q_1 and the rate q_2 at which it leaves. The constant of proportionality is the hydraulic capacitance C.
(c) The rate at which the number N of bacteria in a colony increases with time is proportional to the number of bacteria present at any instant.
(d) The rate at which the number N of radioactive atoms in a sample decreases with time is proportional to the number of radioactive atoms present at any instant.
(e) When a beam of light passes through a block of glass then the intensity I of the transmitted light decreases at a rate proportional to the thickness x of the block.

16 Derive differential equations for the following situations:
(a) A cantilever of negligible mass and length L has a weight W at its free end. Derive a differential equation relating the deflection y of the cantilever with distance x from the fixed end. The bending moment is equal to $EI(\mathrm{d}^2y/\mathrm{d}x^2)$ and in this case is equal to $(WL - Wx)$.
(b) A sphere of radius r and mass m falls through a viscous fluid of density ρ. The forces acting on the sphere are gravity, a buoyant force equal to the weight of fluid displaced by the presence of the sphere (Archimedes' principle) and a resistive force due to the viscosity of the fluid. The resistive force is $6\pi\eta rv$ (Stokes' law), where v is the velocity of the sphere and η the coefficient of viscosity. Derive a differential equation for the velocity in terms of the time.
(c) A charged capacitor C is connected across a resistance R. If the current i from a charged capacitor is $C(\mathrm{d}v_C/\mathrm{d}t)$, where v_C is the potential difference across the capacitor, derive a differential equation relating the potential difference across the capacitor and time.
(d) When a boat, of mass m, switches the engines off it continues to drift with a velocity which decreases with time. If the drag forces acting on the boat are proportional to the velocity v, derive the differential equation relating the velocity and time.

17 For each of the following, verify that the solution is valid for the given differential equation and determine the particular solution which fits the constraints given.

(a) $y = x^2 + A$ for $\dfrac{\mathrm{d}y}{\mathrm{d}x} = 2x$

given $y = 4$ when $x = 0$,

(b) $y = 3x^2 + 2x + A$ for $\frac{dy}{dx} = 6x + 2$

given $y = 5$ when $x = 0$,

(c) $y = A - \cos 3t$ for $\frac{dy}{dt} = 3 \sin 3t$

given $y = 2$ when $t = 0$,

(d) $y = A\,e^{-3t} + B\,e^{4t}$ for $\frac{d^2y}{dt^2} - \frac{dy}{dx} - 12y = 0$

given $y = 10$ when $t = 0$ and $dy/dt = 12$ when $t = 0$.

18 For a cantilever with a uniformly distributed load the differential equation describing how the deflection y of the beam varies with the distance x from the fixed end is

$$\frac{d^2y}{dx^2} = -\frac{w}{2EI}(L^2 - 2Lx + x^2)$$

Verify that the following is a general solution of the above differential equation and determine the constants A and B given that when $x = 0$ both $y = 0$ and $dy/dx = 0$.

$$y = -\frac{w}{2EI}\left(\frac{L^2x^2}{2} - \frac{Lx^3}{3} + \frac{x^4}{12}\right) + Ax + B$$

19 For an electrical circuit consisting of a capacitor in series with a resistor, the differential equation describing how the potential difference v_C varies with time when a voltage of V is suddenly connected to the arrangement is

$$RC\frac{dv_C}{dt} + v_C = V$$

Verify that the general solution of the above equation is

$$v_C = A\,e^{-t/RC} + V$$

and determine the value of the constant A if $v_C = 0$ when $t = 0$.

20 What are the orders and degrees of the following equations and which are linear?

(a) $\frac{dy}{dx} = 2y$, (b) $\frac{d^3y}{dt^3} + 2\frac{d^2y}{dt^2} + 5\frac{dy}{dt} + 3y = 0$,

(c) $\dfrac{d^2y}{dt^2} + 5y^2 = 0$, (d) $y\dfrac{dy}{dt} = 5$, (e) $\dfrac{dy}{dx} + 2\sin 3y = 0$,

(f) $3\dfrac{d^2y}{dt^2} + 5\dfrac{dy}{dt} + 2y = 5\sin 3t$

21 Which of the linear equations in problem 20 are homogeneous?

2 First-order differential equations

2.1 First-order differential equations

There are many situations that can be modelled by first-order differential equations. A common form of first-order differential equation is

$$\frac{dN}{dt} = kN$$

Such an equation states that the rate of change of some quantity N with time t, i.e. dN/dt, is proportional to the amount of that quantity N present at time t. It describes an exponential growth, occurring in such diverse situations as the rate at which money grows when earning compound interest, and the rate at which a population grows when the birth and death rates are constant.

Another form of this equation, which is frequently met in engineering and science, is

$$\frac{dN}{dt} = -kN$$

This describes an exponential decay. Such a relationship exists for such diverse situations as the rate of disintegration of radioactive materials, the rate at which charge flows from a discharging capacitor, and the rate at which the height of water varies with time when it flows out of the base of a container.

The above differential equations are of the form

$$\frac{dy}{dx} = f(y) \qquad [1]$$

Another type of equation considered in this chapter is of the form

$$\frac{dy}{dx} + Py = Q \qquad [2]$$

where P and Q are two functions of x. An example of such a differential equation is the current flow in a series resistor–inductor circuit when a sinusoidal voltage is applied,

$$\frac{di}{dt} + \frac{R}{L}i = \frac{1}{L}\sin \omega t$$

There $P = R/L$ and $Q = (1/L)\sin \omega t$.

In this chapter we will only be considering differential equations which are of the first order and first degree, with chapters 3 and 4 giving examples of applications of such equations in engineering. There are two basic methods that can be used to solve such equations: separation of variables and the use of an integrating factor.

2.2 Separation of variables

There are many first-order differential equations which can be solved by a separation of the variables and then direct integration. Such equations can be of the form:

1 $\dfrac{dy}{dx} = f(x)$

2 $\dfrac{dy}{dx} = f(y)$

3 $f(y)\dfrac{dy}{dx} = f(x)$

4 $\dfrac{dy}{dx} = f(x)f(y)$

5 Equations which can be put into one of the above forms by a suitable change of variable.

The following shows how we can solve each of the above forms of first-order differential equations.

Form 1: Consider first-order differential equations of the form

$$\frac{dy}{dx} = f(x) \qquad [3]$$

We can integrate both sides of the equation with respect to dx.

$$\int \frac{dy}{dx}dx = \int f(x)\, dx$$

This is equivalent to separating the variables x and y in equation [3] to give

$$\mathrm{d}y = f(x)\,\mathrm{d}x$$

and hence

$$\int \mathrm{d}y = \int f(x)\,\mathrm{d}x$$

An example of such an equation is $\mathrm{d}y/\mathrm{d}x = 2x$. This can have the variables separated to give

$$\int \mathrm{d}y = \int 2x\,\mathrm{d}x$$

and so a solution

$$y = x^2 + A$$

where A is the constant of integration. Note that because there are integrations on the left- and right-hand sides of the equation we could have two constants of integration. However, it is usual to combine these into the single constant A.

Form 2: Consider the first-order differential equation

$$\frac{\mathrm{d}y}{\mathrm{d}x} = f(y) \qquad [4]$$

This can be rearranged to give

$$\frac{1}{f(y)}\frac{\mathrm{d}y}{\mathrm{d}x} = 1$$

Integrating both sides of this equation with respect to dx gives

$$\int \frac{1}{f(y)}\frac{\mathrm{d}y}{\mathrm{d}x}\mathrm{d}x = \int 1\,\mathrm{d}x$$

This is equivalent to equation [4] having the variables separated to give

$$\frac{\mathrm{d}y}{f(y)} = \mathrm{d}x$$

and hence

$$\int \frac{\mathrm{d}y}{f(y)} = \int \mathrm{d}x$$

An example of such a differential equation is $\mathrm{d}y/\mathrm{d}x = 4y$. The variables can be separated to give

$$\int \frac{\mathrm{d}y}{y} = \int 4\,\mathrm{d}x$$

and the solution

$$\ln y = 4x + A$$

where A is the constant of integration.

Form 3: Consider a differential equation of the form

$$f(y)\frac{\mathrm{d}y}{\mathrm{d}x} = f(x) \qquad [5]$$

Integrating both sides of this equation with respect to dx gives

$$\int f(y)\frac{\mathrm{d}y}{\mathrm{d}x}\mathrm{d}x = \int f(x)\,\mathrm{d}x$$

This is equivalent to separating the variables to give

$$f(y)\,\mathrm{d}y = f(x)\,\mathrm{d}x$$

with then

$$\int f(y)\,\mathrm{d}y = \int f(x)\,\mathrm{d}x$$

An example of such an equation is $y(\mathrm{d}y/\mathrm{d}x) = 4x$. This can have the variables separated to give

$$\int y\,\mathrm{d}y = \int 4x\,\mathrm{d}x$$

and so the solution

$$\tfrac{1}{2}y^2 = 2x^2 + A$$

where A is the constant of integration.

Form 4: A variation of the form given in equation [5] is

$$\frac{\mathrm{d}y}{\mathrm{d}x} = f(x)f(y) \qquad [6]$$

This can be rearranged to give

$$\frac{1}{f(y)}\frac{\mathrm{d}y}{\mathrm{d}x} = f(x)$$

Integrating both sides of the equation with respect to dx gives

$$\int \frac{1}{f(y)}\frac{\mathrm{d}y}{\mathrm{d}x}\mathrm{d}x = \int f(x)\,\mathrm{d}x$$

This is equivalent to separating the variables to give

$$\frac{1}{f(y)}\mathrm{d}y = f(x)\,\mathrm{d}x$$

and hence

$$\int \frac{1}{f(y)}\mathrm{d}y = \int f(x)\,\mathrm{d}x$$

An example of such a differential equation is $\mathrm{d}y/\mathrm{d}x = 2yx$. The variables can be separated to give

$$\int \frac{\mathrm{d}y}{y} = \int 2x\,\mathrm{d}x$$

and so the solution

$$\ln y = x^2 + A$$

where A is the constant of integration.

Form 5: In some cases the equations may not be separable but can be made separable by an appropriate change of variable. Consider, for example, the differential equation

$$\frac{\mathrm{d}y}{\mathrm{d}x} = \frac{y-x}{y+x}$$

This can be rearranged to give

$$\frac{\mathrm{d}y}{\mathrm{d}x} = \frac{\frac{y}{x} - 1}{\frac{y}{x} + 1}$$

If we now let $v = y/x$ then, since we can write this as $y = vx$, differentiating this with respect to x gives

$$\frac{\mathrm{d}y}{\mathrm{d}x} = v + x\frac{\mathrm{d}v}{\mathrm{d}x} \qquad [7]$$

and so the equation can be written as

$$v + x\frac{dv}{dx} = \frac{v-1}{v+1}$$

$$x\frac{dv}{dx} = \frac{v-1}{v+1} - v = -\left(\frac{v^2+1}{v+1}\right)$$

The variables can now be separated to give

$$\left(\frac{v+1}{v^2+1}\right) dv = -\frac{dx}{x}$$

$$\left(\frac{v}{v^2+1} + \frac{1}{v^2+1}\right) dv = -\frac{dx}{x}$$

Integrating then gives

$$\tfrac{1}{2}\ln(v^2+1) + \tan^{-1} v = -\ln x + A$$

Since $v = y/x$ then the solution is

$$\tfrac{1}{2}\ln\left[\left(\frac{y}{x}\right)^2 + 1\right] + \tan^{-1}\left(\frac{y}{x}\right) = -\ln x + A$$

Example

Solve the following differential equations:

(a) $\frac{dy}{dx} = 2\cos 2x$, (b) $y\frac{dy}{dx} = e^{-x}$, (c) $\frac{dy}{dx} = y^2 x$, (d) $\frac{dy}{dx} = \frac{y}{y+x}$

(a) This differential equation can have the variables separated to give

$$dy = 2\cos 2x\, dx$$

Integration then gives

$$\int dy = \int 2\cos 2x\, dx$$

$$y = \tfrac{2}{2}\sin 2x + A$$

where A is the constant of integration.
(b) This differential equation can have the variables separated to give

$$y\, dy = e^{-x}\, dx$$

Integration then gives

$$\int y\,\mathrm{d}x = \int \mathrm{e}^{-x}\,\mathrm{d}x$$

$$\frac{y^2}{2} = -\mathrm{e}^{-x} + A$$

where A is the constant of integration.

(c) This differential equation can have the variables separated to give

$$\frac{1}{y^2}\,\mathrm{d}y = x\,\mathrm{d}x$$

Integration then gives

$$\int \frac{1}{y^2}\mathrm{d}y = \int x\,\mathrm{d}x$$

$$-\frac{1}{y} = \frac{x^2}{2} + A$$

where A is the constant of integration.

(d) This differential equation can be rearranged to give

$$\frac{\mathrm{d}y}{\mathrm{d}x} = \frac{\frac{y}{x}}{\frac{y}{x} + 1}$$

Let $v = y/x$. Then, using equation [7],

$$\frac{\mathrm{d}y}{\mathrm{d}x} = v + x\frac{\mathrm{d}v}{\mathrm{d}x} = \frac{v}{v+1}$$

$$x\frac{\mathrm{d}v}{\mathrm{d}x} = \frac{v}{v+1} - v = -\frac{v^2}{v+1}$$

$$\left(\frac{v+1}{v^2}\right)\frac{\mathrm{d}v}{\mathrm{d}x} = -\frac{1}{x}$$

The variables can be separated to give

$$\int \left(\frac{1}{v} + \frac{1}{v^2}\right)\mathrm{d}v = -\int \frac{1}{x}\mathrm{d}x$$

and so

$$\ln v - \frac{1}{v} = -\ln x + A$$

Hence the solution is

$$\ln\left(\frac{y}{x}\right) - \frac{x}{y} = -\ln x + A$$

Example

Find the solution of the differential equation $dy/dx = 2y$ given that $y = 4$ when $x = 0$.

The differential equation can have the variables separated to give

$$\frac{dy}{y} = 2\,dx$$

Hence

$$\int \frac{dy}{y} = \int 2\,dx$$

$$\ln y = 2x + A$$

where A is the constant of integration. This can be written as

$$y = e^{2x+A} = e^{A}\,e^{2x} = C\,e^{2x}$$

where C is a constant. Since $y = 4$ when $x = 0$ then $4 = C\,e^{0} = C$. Thus

$$y = 4\,e^{2x}$$

Example

A water tank of constant cross-sectional area 10^6 mm^2 has a hole of cross-sectional area 10^3 mm^2 in its base from which water leaks. Derive a differential equation showing how the height h of water in the tank changes with time and hence determine the time taken to empty the tank if it has an initial depth of 500 mm.

The velocity with which water emerges from the base is given by $v = \sqrt{(2gh)}$. Thus the volume of water leaving the tank per second is $10^3\sqrt{(2gh)}$. This is the change in the volume of water in the tank per second. Thus

$$\frac{d(10^6 h)}{dt} = 10^3\sqrt{2gh}$$

$$\frac{dh}{dt} = 10^{-3}\sqrt{2gh} = 0.140\sqrt{h}$$

In the above, the acceleration due to gravity has been taken as 9.81×10^3 mm/s^2. This is the differential equation describing how the height, in mm, of water in the tank varies with time. This can be solved by separating the variables.

$$\frac{\mathrm{d}h}{\sqrt{h}} = 0.140\,\mathrm{d}t$$

$$\int \frac{\mathrm{d}h}{\sqrt{h}} = \int 0.140\,\mathrm{d}t$$

$$2\sqrt{h} = 0.140t + A$$

where A is the constant of integration. At $t = 0$ then $h = 500$ mm and so

$$2\sqrt{500} = 0 + A$$

Thus $A = 44.7$ and so

$$2\sqrt{h} = 0.140t + 44.7$$

The time taken to empty is the time taken for h to become zero. Thus

$$0 = 0.140t + 44.7$$

and so the time taken is 319 s.

Review problems

1 Find the solutions of the following differential equations:

$$\text{(a) } \frac{\mathrm{d}y}{\mathrm{d}x} = 5x^4,\ \text{(b) } \frac{\mathrm{d}y}{\mathrm{d}x} = 4x,\ \text{(c) } \frac{\mathrm{d}y}{\mathrm{d}x} = \sin x,\ \text{(d) } x\frac{\mathrm{d}y}{\mathrm{d}x} = 5,$$

$$\text{(e) } \frac{\mathrm{d}y}{\mathrm{d}x} = 3 + y,\ \text{(f) } x^2\frac{\mathrm{d}y}{\mathrm{d}x} = x + 1,\ \text{(g) } \frac{\mathrm{d}y}{\mathrm{d}x} = 4\,\mathrm{e}^{2x},$$

$$\text{(h) } (2 + y)\frac{\mathrm{d}y}{\mathrm{d}x} = x^2,\ \text{(i) } \frac{\mathrm{d}y}{\mathrm{d}x} = \left(\frac{y}{x}\right)^2 + \frac{y}{x}$$

2 Find the solutions of the following differential equations which satisfy the conditions given:

(a) $\frac{\mathrm{d}y}{\mathrm{d}x} = 3 - x$ with $y = 5$ when $x = 2$,

(b) $\frac{dy}{dx} = y + 2$ with $y = 0$ when $x = 0$,

(c) $x^2 \frac{dy}{dx} = y$ with $y = 1$ when $x = 1$,

(d) $\frac{dy}{dx} = \frac{3x^2}{2y+1}$ with $y = 0$ when $x = 0$

3 The rate at which a body cools is given by the differential equation

$$\frac{d\theta}{dt} = -k\theta$$

where θ is the excess temperature above the surroundings and k a constant. If the excess temperature at $t = 0$ is θ_0 determine the solution of the equation.

4 For a belt drive, the difference in the tension T between the slack and tight sides of the belt over a pulley is related to the angle of lap θ on the pulley by

$$\frac{dT}{d\theta} = \mu T$$

where μ is the coefficient of friction between the belt and the pulley. Determine the solution of the equation if $T = T_2$ when $\theta = 0°$ and $T = T_1$ when the angle of lap is θ.

2.2.1 Exponential growth and decay

The differential equation

$$\frac{dy}{dx} = ky \qquad [8]$$

where k is a constant, describes an exponential growth. The solution can be obtained by the separation of variables, i.e.

$$\int \frac{dy}{y} = \int k\, dx$$

$$\ln y = kx + A$$

where A is the constant of integration. Taking exponentials

$$y = e^{kx+A}$$

This can be rearranged to give

$$y = e^{A} e^{kx} = C e^{kx} \qquad [9]$$

where $C = e^{A}$ is just another constant. If we have the situation where $y = y_0$ when $x = 0$, then $y_0 = C e^{0} = C$ and equation [9] can be written as

$$y = y_0 e^{kx} \qquad [10]$$

The above equations describe an exponential growth. To describe an exponential decay the differential equation has to be of the form

$$\frac{dy}{dx} = -ky \qquad [11]$$

The variables can be separated to give

$$\int \frac{dy}{y} = -\int k\,dx$$

$$\ln y = -kx + A$$

$$y = e^{-kx+A} = e^{A} e^{-kx} = C e^{-kx} \qquad [12]$$

If we have the situation where $y = y_0$ when $x = 0$, then we have $y_0 = C e^{0} = C$ and so equation [12] can be written as

$$y = y_0 e^{-kx} \qquad [13]$$

Differential equations of the above form occur often in engineering and science. When such equations are met it should be realised that the solution is always of the form e^{ax} multiplied by a constant. It is the initial conditions which enable the constant to be determined.

Example

The rate at which radioactive atoms disintegrate is proportional to the number N of radioactive atoms present, i.e.

$$\frac{dN}{dt} = -kN$$

where k is a constant. (a) Derive an equation indicating how the number of radioactive atoms decreases with time. (b) Derive an equation showing how the activity decreases with time. (c) What will be the time taken for the activity of a sample to decrease by a half, i.e. the so-called half-life?

(a) The variables of the equation can be separated and thus

$$\int \frac{dN}{N} = -\int k\,dt$$

$$\ln N = -kt + A$$

If we take the number of radioactive atoms at time $t = 0$ to be N_0, then

$$N = N_0\,e^{-kt}$$

(b) The activity, i.e. dN/dt, is given by

$$\text{activity} = \frac{dN}{dt} = -kN = -kN_0 e^{-kt}$$

(c) The initial activity is $-kN_0$ and thus

$$\frac{\text{activity}}{\text{initial activity}} = e^{-kt}$$

For the activity to decrease to a half, then the time taken is given by

$$\tfrac{1}{2} = e^{-kt}$$

$$\ln 2 = kt$$

Thus the half-life is $k/\ln 2$.

Review problems

5 The charge q on a discharging capacitor decreases with time according to the differential equation

$$RC\frac{dq}{dt} = -q$$

Solve the equation given that the charge is Q at time $t = 0$.

6 A sum of money deposited in a bank account draws interest of 10% compounded continuously. State the differential equation describing how the sum of money in the account changes with time and its solution.

7 A bullet when travelling with a velocity v is subject to a

retardation which is proportional to v. (a) State the differential equation describing how the velocity varies with time and its solution. (b) If the initial velocity was 200 m/s and this reduced to 80 m/s in 0.5 s, what is the equation relating the velocity and time?

8 A wet sheet loses its moisture at a rate proportional to the moisture content. If a sheet loses half its initial moisture in 50 min, how long will need to elapse before it has lost 80%?

9 Radium has a half-life of 1590 years. What percentage of a sample of radium will radioactively decay in 1 year?

2.3 Integrating factor

Consider the differential equation

$$\frac{\mathrm{d}y}{\mathrm{d}x} = xy$$

We can put this in a suitable form for integration by multiplying both sides of the equation by $1/y$.

$$\frac{1}{y}\frac{\mathrm{d}y}{\mathrm{d}x} = x \qquad [14]$$

which can then be arranged as

$$\int \frac{1}{y}\frac{\mathrm{d}y}{\mathrm{d}x}\,\mathrm{d}x = \int x\,\mathrm{d}x$$

$$\ln y = \tfrac{1}{2}x^2 + A$$

We can integrate each side of the equation with respect to dx because

$$\frac{\mathrm{d}}{\mathrm{d}x}(\ln y) = \frac{1}{y}\frac{\mathrm{d}y}{\mathrm{d}x}$$

$$\frac{\mathrm{d}}{\mathrm{d}x}(\tfrac{1}{2}x^2) = x$$

What has happened is that each side of the equation was multiplied by a factor which resulted in them being put into a form which is recognisable as a derivative and so can be integrated. Such a factor, the $1/y$ in the above example, is called an *integrating factor*.

Consider a first-order differential equation of the form

$$\frac{\mathrm{d}y}{\mathrm{d}x} + Py = Q \qquad [15]$$

where P and Q are functions of x. Such an equation is a *linear differential equation* (see section 1.6.1). We will multiply this equation by some function I of x.

$$I\frac{dy}{dx} + IPy = IQ$$

The aim is to make the left-hand side of the equation into the derivative of some product. Consider the derivative of the product Iy.

$$\frac{d}{dx}(Iy) = I\frac{dy}{dx} + y\frac{dI}{dx}$$

If we want to make the left-hand side of our equation this product then we must have

$$y\frac{dI}{dx} = IPy$$

Separating the variables gives

$$\int\frac{dI}{I} = \int P\,dx$$

$$\ln I = \int P\,dx + A$$

$$I = e^{\int P\,dx + A} = e^{A}\,e^{\int P\,dx}$$

Thus, we have

$$I = C\,e^{\int P(x)\,dx} \qquad [16]$$

where C is a constant and in order to indicate clearly that P is a function of x it has to be written as $P(x)$. This is the integrating factor for the type of differential equation given by [15].

To illustrate the use of this integrating factor, consider the differential equation

$$\frac{dy}{dx} + y = x$$

The function $P(x) = 1$. The integrating factor is thus

$$I = C\,e^{\int P(x)\,dx} = C\,e^{\int 1\,dx} = C\,e^{x}$$

Multiplying both sides of the equation by this integrating factor gives

$$C\,e^x\frac{dy}{dx} + C\,e^x y = C\,e^x x$$

The constant C cancels. We know that the left-hand side of the equation is the derivative of a product Iy, thus the equation can be written as

$$\frac{d}{dx}(y\,e^x) = x\,e^x$$

Hence integrating with respect to x gives

$$y\,e^x = \int x\,e^x\,dx$$

$$y\,e^x = x\,e^x - e^x + A$$

Differential equations of the form

$$\frac{dy}{dx} - ay = b \qquad [17]$$

where a and b are constants are often met in engineering and science. Such an equation has an integrating factor of

$$I = e^{\int -a\,dx} = e^{-ax}$$

and so a general solution of

$$e^{-ax}y = \int e^{-ax}b\,dx = -\frac{b}{a}e^{-ax} + A$$

Thus

$$y = -\frac{b}{a} + Ae^{ax} \qquad [18]$$

An example which gives a general equation of this form is the cooling of an object at a temperature θ in surroundings at a temperature θ_s. Newton's law of cooling gives

$$\frac{d\theta}{dt} = -k(\theta - \theta_s)$$

where k is a constant. This equation can be rearranged to put it in the form of equation [17].

$$\frac{d\theta}{dt} + k\theta = k\theta_s$$

The solution is thus of the same form as equation [18], i.e.

$$\theta = \theta_s + A e^{-kt}$$

Example

Solve the differential equation

$$\frac{dy}{dx} - y = e^{2x}$$

$P(x) = -1$, thus the integrating factor I is given by

$$I = C e^{\int P(x)\,dx} = C e^{\int -1\,dx} = C e^{-x}$$

Hence

$$y e^{-x} = \int e^{-x} e^{2x}\,dx = e^{x} + A$$

$$y = e^{2x} + A e^{x}$$

Example

An electrical circuit has an inductance L in series with a resistance R and is connected to an alternating voltage supply of $V \cos \omega t$. The current i, at a time t after the switch is closed and the voltage is applied, is given by the differential equation

$$L\frac{di}{dt} + Ri = V \cos \omega t$$

Solve the equation if the current is zero when $t = 0$.

The equation can be rearranged to put it in the form of equation [15].

$$\frac{di}{dt} + \frac{R}{L} i = \frac{V}{L} \cos \omega t$$

The integrating factor is then given by equation [16] as

$$I = C e^{\int P(t)\,dt} = C e^{\int (R/L)\,dt} = C e^{Rt/L}$$

On multiplying both sides of the equation by this integrating factor the constant C can be cancelled. Thus

$$\frac{d}{dt}(i\, e^{Rt/L}) = e^{Rt/L} \frac{V}{L} \cos \omega t$$

Integrating both sides of the equation with respect to dt and using integration by parts, i.e. $\int u\,\mathrm{d}v = uv - \int v\,\mathrm{d}u$, then gives

$$i\,\mathrm{e}^{Rt/L} = \int \mathrm{e}^{Rt/L}\frac{V}{L}\cos\omega t\,\mathrm{d}t$$

$$= \frac{V\,\mathrm{e}^{Rt/L}}{R^2+\omega^2L^2}(R\cos\omega t + L\omega\sin\omega t) + A$$

Since $i = 0$ when $t = 0$ then

$$0 = \frac{V}{R^2+\omega^2L^2}R + A$$

$$A = -\frac{VR}{R^2+\omega^2L^2}$$

Thus

$$i\,\mathrm{e}^{Rt/L} = \frac{V\,\mathrm{e}^{Rt/L}}{R^2+\omega^2L^2}(R\cos\omega t + L\omega\sin\omega t) - \frac{VR}{R^2+\omega^2L^2}$$

$$i = \frac{V}{R^2+\omega^2L^2}(R\cos\omega t + L\omega\sin\omega t - R\,\mathrm{e}^{-Rt/L})$$

Review problems

10 Solve the following differential equations:

(a) $\dfrac{\mathrm{d}y}{\mathrm{d}x} + 3y = 6$, (b) $\dfrac{\mathrm{d}y}{\mathrm{d}x} + 2xy = x$, (c) $\dfrac{\mathrm{d}y}{\mathrm{d}x} - 3y = \mathrm{e}^{2x}$,

(d) $\dfrac{\mathrm{d}y}{\mathrm{d}x} + 3y = x$ with $y = 1$ when $x = 0$,

(e) $\dfrac{\mathrm{d}y}{\mathrm{d}x} + y = 2$ with $y = 0$ when $x = 1$

2.4 Complementary function and particular integral

Suppose we have the non-homogeneous equation

$$\frac{\mathrm{d}y}{\mathrm{d}x} + y = 5$$

This differential equation has the solution

$$y = C\,\mathrm{e}^{-x} + 5$$

Now suppose we have a corresponding homogeneous equation, i.e. the equation with the right-hand side put equal to zero,

$$\frac{dy}{dx} + y = 0$$

This has the solution

$$y = C e^{-x}$$

Thus the solution of the non-homogeneous equation is equal to the sum of the solution of the homogeneous equation, called the *complementary function*, plus another term, called the *particular integral*. The particular integral is any solution of the non-homogeneous equation. Thus, in the above case the particular integral solution is for $y = 5$. Since dy/dx of 5 is 0 then substituting for y in the equation gives $0 + 5 = 5$ and confirms that it is indeed a solution.

Thus, if we consider the general form of a linear non-homogeneous differential equation

$$\frac{dy}{dx} + P(x)y = Q(x)$$

then we can write the solution in the form

$$y = y_c + y_p$$

where y_c is the complementary function and y_p the particular integral. The complementary function is the solution of the corresponding homogeneous equation, and the particular integral can be determined by educated guesswork and then tried to check (see chapter 10 for another way of determining the particular integral). The usual method is to try a function of the same form as that on the right-hand side of the equation. Thus, if the right-hand term is a constant then we can try $y = A$, where A is a constant. If it is an exponential function then we can try $y = A e^{kx}$, where A and k are constants. If it is of the form $a + bx + cx^2 + \ldots$ then we can try $y = A + Bx + Cx^2 + \ldots$ If it is a sine or cosine then we can try $y = A \sin \omega x + B \cos \omega x$. If it is a sum of some of these terms above then we can try the sum of the corresponding terms indicated, if a product then we try the corresponding product.

To illustrate the above, consider the differential equation

$$\frac{dy}{dx} + 2y = 4x$$

The corresponding homogeneous equation is

$$\frac{dy}{dx} + 2y = 0 \text{ or } \frac{dy}{dx} = -2y$$

This is the form of differential equation that occurs with an exponential decay. Thus the solution will be of the form $y = C\mathrm{e}^{-kx}$. If we try this as the solution then

$$-kC\mathrm{e}^{-kx} = -2C\mathrm{e}^{-kx}$$

and so $k = 2$. The complementary solution is thus

$$y_c = C\mathrm{e}^{-2x}$$

Because the term to the right of the equals sign is $4x$ the particular integral we can try is $y = A + Bx$. The non-homogeneous equation then becomes, with this solution,

$$B + 2(A + Bx) = 4x$$

Hence, equating coefficients of x gives $B = 2$ and equating constant terms gives $B + 2A = 0$ and so $A = -1$. Thus

$$y_p = -1 + 2x$$

Thus the solution of the non-homogeneous equation is

$$y = y_c + y_p = C\mathrm{e}^{-2x} - 1 + 2x$$

Example

Determine the solution of the differential equation

$$\frac{dy}{dx} + 2y = x^2$$

The corresponding homogeneous equation is

$$\frac{dy}{dx} + 2y = 0$$

and, as already indicated above, gives the complementary function as

$$y_c = C\mathrm{e}^{-2x}$$

For the particular integral we can try $y = A + Bx + Cx^2$. Using this value in the non-homogeneous equation gives

$$B + 2Cx + 2A + 2Bx + 2Cx^2 = x^2$$

Equating coefficients of x^2 gives $C = \frac{1}{2}$. Equating coefficients of x gives $B = -\frac{1}{2}$. Equating constants gives $A = \frac{1}{4}$. Thus

$$y_p = \tfrac{1}{4} - \tfrac{1}{2}x + \tfrac{1}{2}x^2$$

and hence

$$y = y_c + y_p = C\,e^{-2x} + \tfrac{1}{4} - \tfrac{1}{2}x + \tfrac{1}{2}x^2$$

Review problems

11 Determine the solutions of the following differential equations:

(a) $\dfrac{dy}{dx} - y = 4\,e^{3x}$, (b) $\dfrac{dy}{dx} - y = 4$, (c) $\dfrac{dy}{dx} - y = 4x + 2$,

(d) $\dfrac{dy}{dx} - y = 4\sin 2x$

Further problems

12 Solve the following differential equations using the separation of variables method:

(a) $(y+5)\dfrac{dy}{dx} = x^2$, (b) $\dfrac{dy}{dx} = e^x$, (c) $x\dfrac{dy}{dx} = 1 - x^2$,

(d) $\dfrac{dy}{dx} = \sin 2x$, (e) $\dfrac{dy}{dx} + 2xy = 0$, (f) $\dfrac{dy}{dx} - \sin x = 2x$,

(g) $y\dfrac{dy}{dx} = 2 - y^2$, (h) $\dfrac{dy}{dx} = y\cos x$, (i) $\dfrac{dy}{dx} = \dfrac{y^2 - x^2}{3xy}$,

(j) $\dfrac{dy}{dx} = \dfrac{y+x}{2x}$

13 Solve the following differential equations, using the separation of variables method, for the conditions given:

(a) $\dfrac{dy}{dx} = 2y - 1$ with $y = 1$ when $x = 1$,

(b) $\dfrac{dy}{dx} = y$ with $y = 2$ when $x = 0$,

(c) $\dfrac{dy}{dx} = 6x^2 e^{-y}$ with $y = 2$ when $x = 0$,

(d) $\frac{dy}{dx} = 10x^4$ with $y = 0$ when $x = 0$,

(e) $2\frac{dy}{dx} = \frac{1}{\cos 2y}$ with $y = 0$ when $x = 2$,

(f) $2\frac{dy}{dx} + \sin 2x = 0$ with $y = 2$ when $x = \frac{\pi}{4}$,

(g) $\frac{dy}{dx} = \frac{e^x}{e^{2y}}$ with $y = 0$ when $x = 0$,

(h) $\frac{dy}{dx} = 3yx^2$ with $y = 1$ when $x = 0$,

(i) $x^2\frac{dy}{dx} = y^2 + xy$ with $y = 1$ when $x = 1$.

14 The rate of change of resistance R with temperature θ for a metal is given by

$$\frac{dR}{d\theta} = \alpha R$$

where α is the temperature coefficient of resistance for the metal. If $R = R_0$ when $\theta = 0$, determine the solution to the equation.

15 If a quarter of an additive dissolves in a liquid in 60 s, how long will is take for half of it to dissolve if the rate at which the additive dissolves is given by

$$\frac{dA}{dt} = -kA$$

where A is the amount of additive remaining undissolved after a time t?

16 The intensity of light I transmitted through a distance x of water is determined by the differential equation

$$\frac{dI}{dx} = -4I$$

After what distance of water will the intensity have been reduced to a half?

17 Solve the following differential equations using the integrating factor method:

(a) $\frac{dy}{dx} - y = 2x$, (b) $2x\frac{dy}{dx} + y = 4x^2$,

(c) $(1+x^2)\frac{dy}{dx}+3xy=3x$, (d) $\frac{dy}{dx}-2y=4x$,

(e) $\frac{dy}{dx}+\frac{y}{x^2-x}=\frac{x+1}{x^2-x}$, (f) $\frac{dy}{dx}+3y=4$,

(g) $x\frac{dy}{dx}-4y=4x^4$, (h) $\frac{dy}{dx}+2y=e^x$, (i) $\frac{dy}{dx}+yx+2x=0$

18 An electrical circuit has a resistance R in series with an inductance L. At time $t = 0$ a ramp voltage of 1 V/s is applied. The differential equation describing how the circuit current varies with time is

$$L\frac{di}{dt}+Ri=1t$$

Solve the equation, taking the current to be zero at $t = 0$.

19 An electrical circuit consists of a capacitor C in series with a resistance R. At time $t = 0$ the capacitor is uncharged and the current in the circuit is zero. Then a switch is closed and a constant voltage V applied to the circuit. The differential equation describing how the potential difference v across the capacitor changes with time is

$$RC\frac{dv}{dt}+v=V$$

Solve the equation.

20 In a chemical reaction a substance A changes into substance B. If a was the initial mass of A, then the mass x of B present after a time t is given by the differential equation

$$\frac{dx}{dt}=k(a-x)$$

where k is a constant. Solve the equation, given that $x = 0$ when $t = 0$.

21 In a chemical reaction a substance A reacts with a substance B to give a substance C. The differential equation describing how the amount x of substance C changes with time is given by

$$\frac{dx}{dt}=k(a-x)(b-x)$$

where k is a constant, a the amount of A at a time t and b the amount of B at that time. Solve the equation given that $x = 0$ at $t = 0$.

22 By determining the complementary functions and the particular

integrals, obtain solutions for the following differential equations:

$$\text{(a) } \frac{dy}{dx} + 2y = 2x, \text{ (b) } \frac{dy}{dx} + 2y = e^x,$$

$$\text{(c) } \frac{dy}{dx} + 2y = 2x - 7, \text{ (d) } \frac{dy}{dx} + 2y = x^2$$

3 First-order differential equations: electric circuits

3.1 Transients

This chapter illustrates the use of first-order differential equations to model the transient currents and voltages that occur with electrical circuits involving resistors and capacitors or resistors and inductors. When a voltage is applied, it takes a little time for the current and voltage to settle down to their *steady state value*. The currents and voltages during this settling down time are referred to as *transients*. Chapter 9 is a consideration of transient currents and voltages with circuits involving resistors, capacitors and inductors.

A common form of voltage input to a circuit is that which occurs when a constant voltage is suddenly switched into a circuit. A sudden jump in voltage is referred to as a *step voltage* since the voltage steps abruptly up from zero to some value and then remains constant at that value. Figure 3.1 illustrates such a voltage for a step of size V.

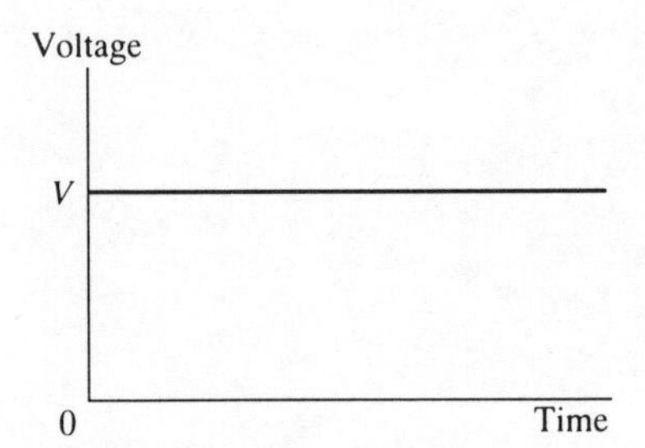

Fig. 3.1 A step voltage

3.2 Circuit components

Consider the relationships between the current through a component and the potential difference across it. When a pure *resistor*, i.e. a component having only resistance, has a potential difference v applied across it then the current i through it is given by *Ohm's law* as

$$i = \frac{v}{R} \qquad [1]$$

When a pure *capacitor*, i.e. a component having only capacitance, has a potential difference v applied across it then the charge q that occurs on the capacitor plates is given by

$$q = Cv \qquad [2]$$

where C is the capacitance. Since the current i is the rate of movement of charge, then

$$i = \frac{dq}{dt} = C\frac{dv}{dt} \qquad [3]$$

The current is thus proportional to the rate of change of the potential difference across the capacitor.

It is sometimes more convenient to have this equation in another form. Equation [3] can be rearranged to give

$$dv = \frac{1}{C} i \, dt$$

and hence, if $v = 0$ when $t = 0$ and is v at time t,

$$\int_0^v dv = \frac{1}{C} \int_0^t i \, dt$$

Thus

$$v = \frac{1}{C} \int_0^t i \, dt \qquad [4]$$

When a pure *inductor*, i.e. a component having only inductance, has a current flowing through it then an induced e.m.f. is produced in the component. The induced e.m.f. is proportional to the rate of change of the current, the constant of proportionality being called the *inductance* L, i.e.

$$\text{induced e.m.f.} = -L\frac{di}{dt} \qquad [5]$$

If the component has only inductance and no resistance, then there can be no potential drop across the component due to its resistance. Thus to maintain the current through the inductor, the voltage source supplying the current must supply a potential difference v which just cancels out the induced e.m.f. Thus the potential difference supplied by the source is

$$v = L\frac{di}{dt} \qquad [6]$$

The potential difference across the inductor is thus proportional to the rate of change of current through it.

It is sometimes more convenient to have this equation in another form. Equation [6] can be rearranged to give

$$di = \frac{1}{L} v \, dt$$

Hence if $v = 0$ when $t = 0$ and is v at time t,

$$\int_0^i \mathrm{d}i = \frac{1}{L}\int_0^t v\,\mathrm{d}t$$

and so

$$i = \frac{1}{L}\int_0^t v\,\mathrm{d}t \qquad [7]$$

3.3 *RC* circuits

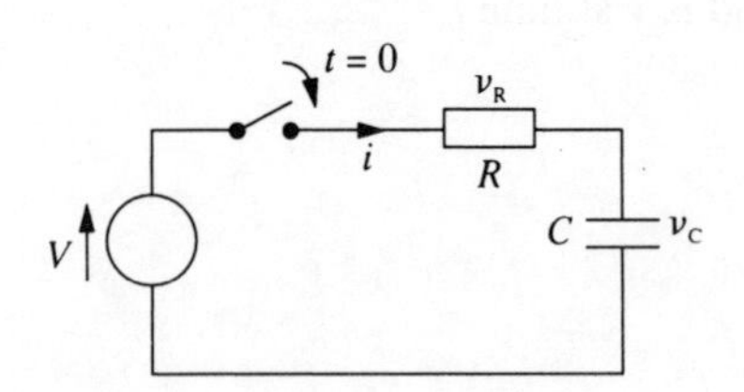

Fig. 3.2 Series *RC* circuit

Consider the circuit shown in figure 3.2. It consists of a capacitor C in series with a resistor R, a constant voltage V being switched into the circuit at time $t = 0$. Applying Kirchhoff's voltage law to the circuit

$$v_R + v_C = V$$

where v_R is the potential difference across the resistor at some instant of time and v_C is that across the capacitor. But $v_R = Ri$ (equation [1]), thus

$$Ri + v_C = V$$

Since $i = C\,\mathrm{d}v_C/\mathrm{d}t$ (equation [3]) then

$$RC\frac{\mathrm{d}v_C}{\mathrm{d}t} + v_C = V \qquad [8]$$

This is a first-order, linear, differential equation which describes the relationship between the potential difference across the capacitor and time when a step voltage V is applied.

Example

A circuit of the form shown in figure 3.2 has a capacitor of 8 μF in series with a resistance of 1 MΩ. Write the differential equation relating the voltage across the capacitor with time when a constant voltage of 10 V is suddenly switched into the circuit.

Equation [8], with the above values, becomes

$$1 \times 10^6 \times 8 \times 10^{-6}\frac{\mathrm{d}v_C}{\mathrm{d}t} + v_C = 10$$

$$8\frac{\mathrm{d}v_C}{\mathrm{d}t} + v_C = 10$$

Review problems

1 A resistance of 2 MΩ is connected in series with a 2 μF capacitor, as in the circuit shown in figure 3.2. Write the differential equation relating the potential difference across the capacitor with time when a constant voltage of 5 V is suddenly applied to the circuit.

3.3.1 Capacitor discharge

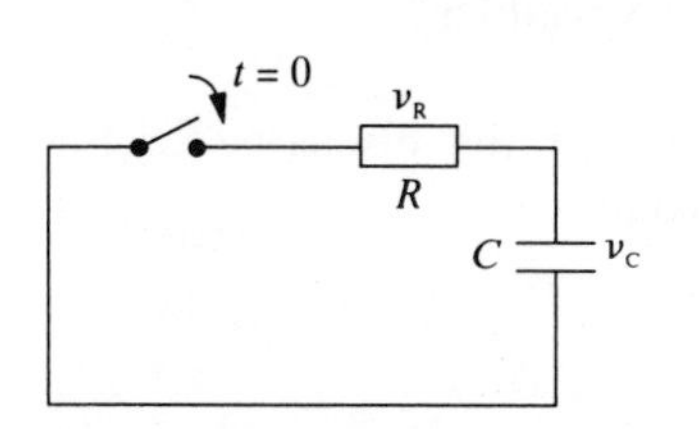

Fig. 3.3 Discharge circuit

Consider the circuit shown in figure 3.3. A charged capacitor is discharged through a resistor. Applying Kirchhoff's voltage law gives

$$v_R + v_C = 0$$

Since $v_R = Ri$ (equation [1]), then

$$Ri + v_C = 0$$

But $v_C = C\, dv_C/dt$ (equation [3]) and so

$$RC\frac{dv_C}{dt} + v_C = 0 \qquad [9]$$

This is a first-order, linear, differential equation describing how the potential difference across the capacitor varies with time when a charged capacitor is allowed to discharge through a resistance.

Note that the difference between the differential equations describing the charging and discharging of a capacitor, i.e. equations [8] and [9], is just the right-hand side of the equation. In the case of the discharging circuit the right-hand side is zero, since there is no voltage source in the circuit, while for the charging circuit when a constant voltage V is applied the right-hand side is just the input voltage source V. If the charging equation had been derived for other forms of input voltage, e.g. a ramp input voltage, then the right-hand side would have been just the function describing the input voltage.

Review problems

2 A charged 4 μF capacitor is allowed to discharge through a resistance of 1 MΩ. Write a differential equation relating the potential difference across the capacitor with time.

3.3.2 Solving the differential equations

Such differential equations can be solved by the separation of variables or by the use of integrating factors (see chapter 2 for details of these methods). A technique that is however commonly used, with both the first-order circuits in this chapter and the second-order circuits in chapter 9, is to recognise from the form of the differential equation the type of general solution that would fit the equation and then establish that such a solution is valid. Section 3.3.3 shows how the equations can be solved using the separation of variables and then section 3.3.4 the technique of assuming a general solution.

3.3.3 Solving using the separation of variables

Consider the differential equation [9] for the discharge of the charged capacitor through a resistor (figure 3.3).

$$RC\frac{\mathrm{d}v_C}{\mathrm{d}t} + v_C = 0$$

This equation can be rearranged to separate the variables and give

$$\frac{\mathrm{d}v_C}{v_C} = -\frac{1}{RC}\mathrm{d}t$$

Integrating both sides of the equation gives

$$\int \frac{\mathrm{d}v_C}{v_C} = -\frac{1}{RC}\int \mathrm{d}t$$

$$\ln v_C = -\frac{1}{RC}t + A$$

where A is the constant of integration. This is the general solution. We can obtain a particular solution by taking into account the constraints. Thus, since at time $t = 0$ the capacitor is fully charged, say to voltage V_0, then $A = \ln V_0$ and so

$$\ln v_C = -\frac{1}{RC}t + \ln V_0$$

$$\ln \frac{v_C}{V_0} = -\frac{1}{RC}t$$

$$\frac{v_C}{V_0} = \mathrm{e}^{-t/RC}$$

$$v_C = V_0\mathrm{e}^{-t/RC} \qquad [10]$$

The potential difference across the resistor v_R is, since $v_R + v_C = 0$,

$$v_R = -v_C = -V_0\mathrm{e}^{-t/RC} \qquad [11]$$

The circuit current i is given by Ohm's law as v_R/R and so is

$$i = -\frac{V_0}{R}e^{-t/RC} \qquad [12]$$

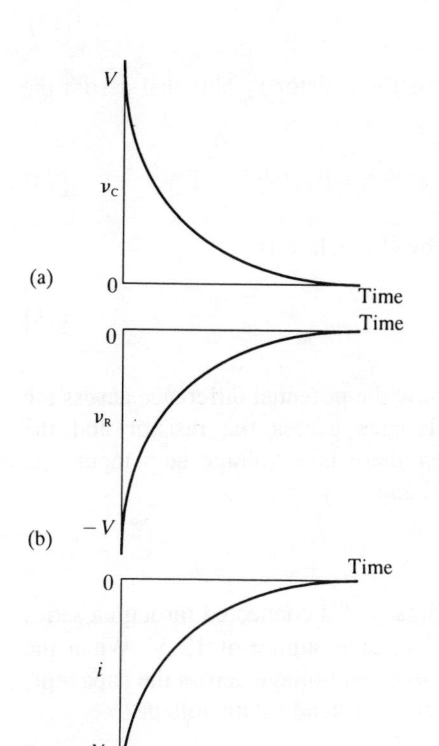

Fig. 3.4 Discharge of a capacitor through a resistance

Figure 3.4 shows graphs of how the potential difference across the capacitor, the potential difference across the resistor and the circuit current vary with time, i.e. graphs of equations [10], [11] and [12].

Now consider the differential equation [8] which occurs when a step voltage V is applied to the RC circuit.

$$RC\frac{dv_C}{dt} + v_C = V$$

This differential equation differs only from the equation occurring when we just have a capacitor discharging through a resistor, equation [8], by including the term V. This is the voltage which is applied to the circuit. With the discharge circuit, this term has the value 0 since there is no voltage input to the circuit. The differential equation with zero voltage input is said to be the *natural equation* since it is the equation which occurs for the circuit when it is all alone with no external input. It is the source-free equation. When a voltage input term is included then the equation is said to include a *forcing function*, the V in this case.

The differential equation with a forcing function can be rearranged to give

$$\frac{dv_C}{v_C - V} = -\frac{1}{RC}dt$$

Integrating both sides of the equation gives

$$\int \frac{dv_C}{v_C - V} = -\frac{1}{RC}\int dt$$

$$\ln(v_C - V) = -\frac{1}{RC}t + A$$

where A is the constant of integration. This is the general solution. To obtain the particular solution we need to take into account the constraints on the values of v_C and t. This equation can be re-arranged to become

$$v_C - V = e^{-t/RC}\, e^{A}$$

Since $v_C = 0$ when $t = 0$, then

$$-V = 1 \times e^{A}$$

Thus

$$v_C - V = -V e^{-t/RC}$$

$$v_C = V(1 - e^{-t/RC}) \qquad [13]$$

The potential difference across the resistor v_R plus that across the capacitor v_C equals V, thus

$$v_R = V - v_C = V - V(1 - e^{-t/RC}) = V e^{-t/RC} \qquad [14]$$

The circuit current i is given by Ohm's law as

$$i = \frac{v_R}{R} = \frac{V}{R} e^{-t/RC} \qquad [15]$$

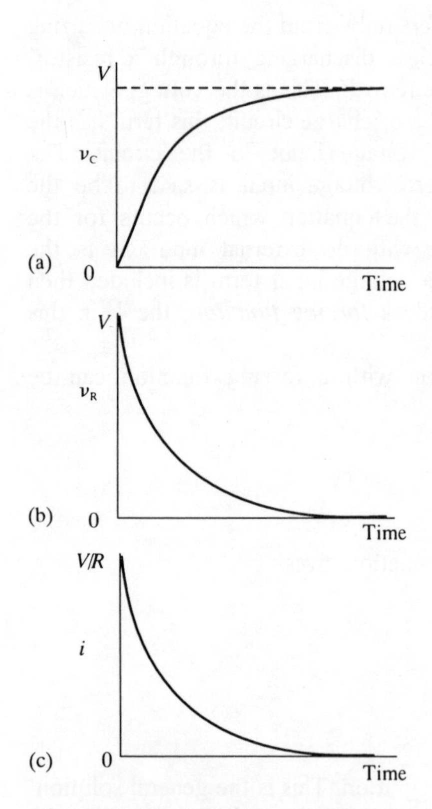

Fig. 3.5 Response of an *RC* circuit to a step input

Figure 3.5 shows graphs of how the potential difference across the capacitor, the potential difference across the resistor and the current vary with time when there is a voltage step input, i.e. graphs of equations [13], [14] and [15].

Example

An initially uncharged 4 μF capacitor is connected through a series 1 MΩ resistor to a switched voltage supply of 12 V. When the switch is closed what is (a) the initial voltage across the capacitor, (b) the voltage after 2 s, (c) the final steady state voltage?

This circuit is of the form shown in figure 3.2 and will consequently have a differential equation of the form shown in equation [8]. Thus

$$RC\frac{dv_C}{dt} + v_C = V$$

$$4\frac{dv_C}{dt} + v_C = 12$$

The solution of this equation will be of the form shown in equation [13]. Thus

$$v_C = V(1 - e^{-t/RC})$$

$$v_C = 12(1 - e^{-t/4})$$

(a) Initially, with $t = 0$, then $v_C = 0$.
(b) After 2 s then

$$v_C = 12(1 - e^{-2/4}) = 4.72 \text{ V}$$

(c) After an infinite time then $v_C = 12$ V.

Review problems

3 An initially uncharged 8 μF capacitor is connected in series to a 1 MΩ resistor. A voltage supply of 20 V is connected to the circuit. After 4 s what will be (a) the potential difference across the capacitor, (b) the current, and (c) the potential difference across the resistor?

4 A 1000 μF capacitor has been charged to a potential difference of 12 V. The capacitor is then discharged through a 20 kΩ resistor. After 2 s, what will be (a) the potential difference across the capacitor, (b) the circuit current?

3.3.4 Assuming a general solution

For the discharge of a charged capacitor through a resistance, i.e. the circuit shown in figure 3.3, the potential difference across the capacitor is described by the differential equation [9]

$$RC\frac{dv_C}{dt} + v_C = 0$$

We can write this in the form

$$\frac{dv_C}{dt} = -\frac{1}{RC}v_C$$

This is a differential equation of the form described in chapter 2 for the exponential decay of a quantity. Thus for this differential equation we can try a general solution of the form

$$v_C = A\,e^{st}$$

where A and s are constants. With this solution

$$\frac{dv_C}{dt} = As\,e^{st}$$

Thus the differential equation becomes

$$RCAs\,e^{st} + A\,e^{st} = 0$$

and so we must have $s = -1/RC$. Thus the general solution of the differential equation is

$$v_C = A\,e^{-t/RC} \qquad [16]$$

The particular solution can be found from the constraint that we have $v_C = V_0$ at $t = 0$. Thus $A = V_0$. Hence

$$v_C = V_0 \, e^{-t/RC} \qquad [17]$$

Since $v_C + v_R = 0$, then

$$v_R = -V_0 \, e^{-t/RC} \qquad [18]$$

The circuit current i is given by Ohm's law as

$$i = \frac{v_R}{R} = -\frac{V_0}{R} e^{-t/RC} \qquad [19]$$

Now consider the circuit, figure 3.2, where a step voltage is applied to a capacitor in series with a resistor. The relationship between the potential difference across the capacitor and time is described by the differential equation [8] as

$$RC\frac{dv_C}{dt} + v_C = V$$

Because this equation is linear (see chapter 1 for a discussion of linear differential equations and chapter 2 for a discussion of the particular integrals and complementary solutions) then the solution can be expressed in the form

$$v_C = v_n + v_f \qquad [20]$$

This gives

$$\frac{dv_C}{dt} = \frac{dv_n}{dt} + \frac{dv_f}{dt}$$

and so the differential equation can be written as

$$RC\left(\frac{dv_n}{dt} + \frac{dv_f}{dt}\right) + v_n + v_f = V$$

This gives, on rearrangement,

$$\left(RC\frac{dv_n}{dt} + v_n\right) + \left(RC\frac{dv_f}{dt} + v_f\right) = V$$

Hence for the complementary solution we have

$$RC\frac{dv_n}{dt} + v_n = 0 \qquad [21]$$

and for the particular integral,

$$RC\frac{dv_f}{dt} + v_f = V \qquad [22]$$

Equation [21] is the same form as the discharge equation [9], hence the subscript n to indicate that v_n is the *natural function.* For this differential equation we can, as earlier in this section, try a solution of the form

$$v_f = A\,e^{st}$$

Substituting this into the differential equation [21] gives

$$RCAs\,e^{st} + A\,e^{st} = 0$$

and so $s = -1/RC$. Thus the general solution is

$$v_n = A\,e^{-t/RC} \qquad [23]$$

Equation [22] contains the input function V and so contains the *forcing function*, hence the subscript f. The input function is a step for which for $t > 0$ we have a constant value of the input voltage. Hence, for this forcing differential equation we try a solution of the form (see chapter 2 for a discussion of the particular integral)

$$v_f = k$$

where k is a constant. Thus we have $dk/dt = 0$ and so can write for the differential equation [22]

$$0 + k = V$$

and so the general solution is

$$v_f = V$$

The general solution of the differential equation [8] is thus given by equation [20] as

$$v_C = A\,e^{-t/RC} + V$$

The particular solution can be obtained by using the constraint that $v_C = 0$ when $t = 0$. Thus

$$0 = A\,e^{0} + V$$

and so $A = -V$. The particular solution is thus

$$v_C = -Ve^{-t/RC} + V = V(1 - e^{-t/RC}) \quad [24]$$

Since $v_R + v_C = V$ then the potential difference across the resistor is

$$v_R = V - v_C = Ve^{-t/RC} \quad [25]$$

The circuit current is given by Ohm's law as

$$i = \frac{v_R}{R} = \frac{V}{R}e^{-t/RC} \quad [26]$$

Example

An initially uncharged 8 μF capacitor is connected through a series 1 MΩ resistor to a voltage supply of 20 V. State the differential equation describing how the potential difference across the capacitor varies with time after the voltage supply is connected, solve the equation and hence determine the potential difference 2 s after the connection of the voltage.

The differential equation will be of the form described in equation [8], i.e.

$$RC\frac{dv_C}{dt} + v_C = V$$

and so

$$8\frac{dv_C}{dt} + v_C = 20$$

The solution to this equation can be obtained as described above. Thus if we take $v_C = v_n + v_f$, then we can write the differential equation as the sum of two other differential equations. Thus the equation which involves just v_n and will give the complementary solution is

$$8\frac{dv_n}{dt} + v_n = 0$$

and the equation which involves just v_f and will give the particular integral solution is

$$8\frac{dv_f}{dt} + v_f = 20$$

The complementary solution equation describes an exponential decay and thus has a solution of the form

$$v_n = A\,e^{st}$$

Hence, using this value, the equation becomes

$$8As\,e^{st} + A\,e^{st} = 0$$

Thus we have $s = -1/8$ and so $v_n = Ae^{-t/8}$. The equation giving the particular integral solution can be considered to have a solution of the form $v_f = k$. Hence, using this value, the differential equation becomes

$$0 + k = 20$$

and thus the particular integral is $v_f = 20$. The solution is thus

$$v_C = v_n + v_f = A\,e^{-t/8} + 20$$

The capacitor is initially uncharged and so $v_C = 0$ when $t = 0$. Thus

$$0 = A + 20$$

Hence the solution is

$$v_C = -20\,e^{-t/8} + 20 = 20(1 - e^{-t/8})$$

For $t = 2$ s then

$$v_C = 20(1 - e^{-2/8}) = 4.42\text{ V}$$

Example

Determine the differential equation and its solution for the potential difference across the initially uncharged capacitor in the circuit shown in figure 3.6. The voltage input to the circuit, which is connected at time $t = 0$, is a ramp voltage of $4t$ V (such a voltage increases at the rate of 4 V every 1 s).

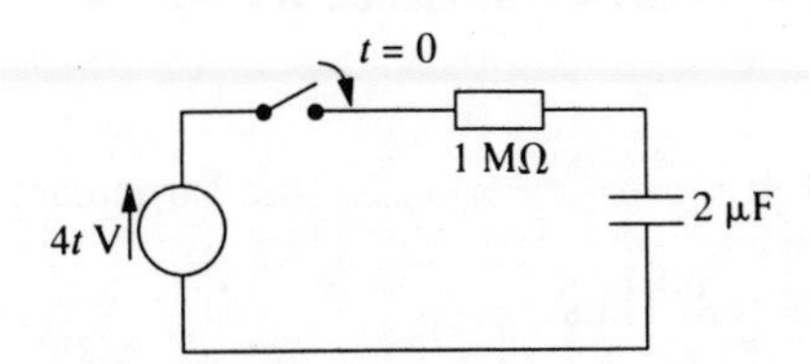

Fig. 3.6 Example

Applying Kirchhoff's voltage law to the circuit

$$v_R + v_C = 4t$$

where v_R is the potential difference across the resistor at some instant of time and v_C is that across the capacitor. But $v_R = Ri$ (equation [1]), thus

$$Ri + v_C = 4t$$

Since $i = C\,dv_C/dt$ (equation [3]) then

$$RC\frac{dv_C}{dt} + v_C = 4t$$

$$2\frac{dv_C}{dt} + v_C = 4t$$

Taking $v_C = v_n + v_f$, we can write the above differential equation as the sum of two differential equations, one of which yields the complementary solution and the other the particular integral, i.e.

$$2\frac{dv_n}{dt} + v_n = 0$$

and

$$2\frac{dv_f}{dt} + v_f = 4t$$

For the natural response, since the equation describes an exponential decay then the solution is of the form

$$v_n = A\,e^{st}$$

Hence, using this value in the differential equation gives

$$2As\,e^{st} + A\,e^{st} = 0$$

Thus $s = -1/2$ and so $v_n = A\,e^{-t/2}$.

For the forced response, since the right-hand side of the equation is $4t$ we can try a solution of the form $v_f = A + Bt$ (see chapter 2 and the discussion of the form of the particular integral). The differential equation thus becomes

$$2B + A + Bt = 4t$$

Thus we must have $B = 4$ and $A = -2B = -8$. Hence $v_f = -8 + 4t$ and so

$$v_C = v_n + v_f = A\,e^{-t/2} - 8 + 4t$$

Since $v_C = 0$ when $t = 0$, then $A = 8$. Thus

$$v_C = 8\,e^{-t/2} - 8 + 4t$$

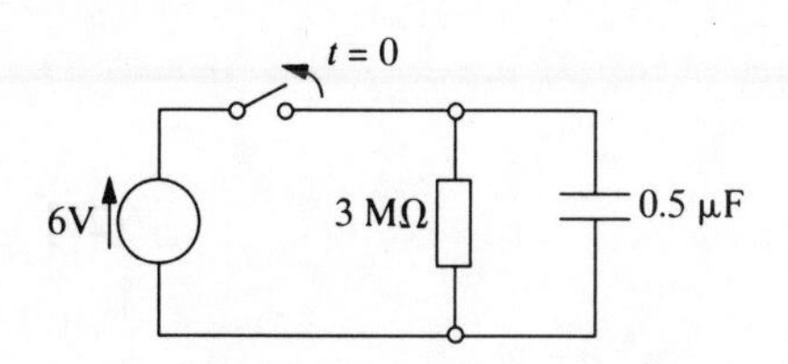

Fig. 3.7 Problem 5

Review problems

5 Determine the differential equation and its solution for the potential difference across the capacitor in the circuit shown in figure 3.7.

6 Determine the differential equation and its solution for the potential difference across the capacitor in a circuit consisting of a 4 MΩ resistor in series with a 3 μF capacitor, initially uncharged, when a ramp voltage of $2t$ is connected to the arrangement at time $t = 0$.

7 Repeat problem 6 for the condition that the capacitor is not uncharged at time $t = 0$ but is charged to a potential difference of 6 V.

3.4 *RL* circuits

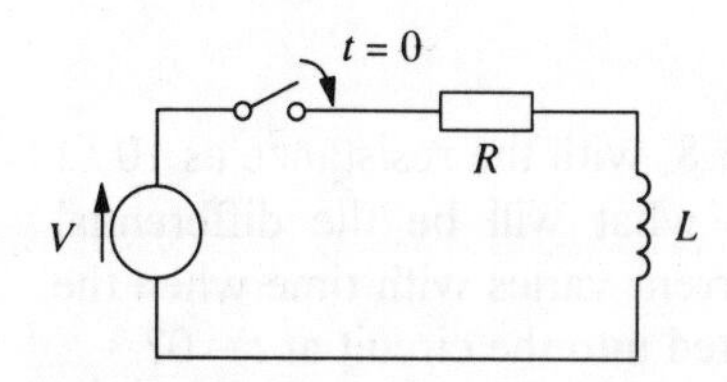

Fig. 3.8 Series *RL* circuit

Consider the circuit shown in figure 3.8 of a resistance in series with an inductance. When the switch is closed and the constant voltage V is applied then

$$v_L + v_R = V$$

where v_R is the potential difference across the resistor, and given by $v_R = Ri$, and v_L is the potential difference across the inductance. Thus, since $v_L = L\, \mathrm{d}i/\mathrm{d}t$ then

$$L\frac{\mathrm{d}i}{\mathrm{d}t} + Ri = V \qquad [27]$$

This first-order equation describes how the current in the circuit changes with time. The steady state current I will be attained when the current ceases to change with time, i.e. $\mathrm{d}i/\mathrm{d}t = 0$. Thus $V = RI$. Hence equation [27] can be written as

$$L\frac{\mathrm{d}i}{\mathrm{d}t} + Ri = RI$$

$$\frac{L}{R}\frac{\mathrm{d}i}{\mathrm{d}t} + i = I \qquad [28]$$

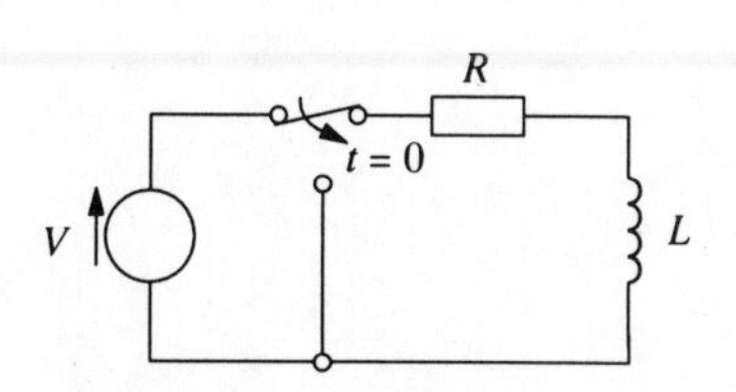

Fig. 3.9 Series *RL* circuit

Consider, after a steady current has been flowing through a series *RL* circuit, the applied voltage is removed and the circuit short-circuited, as in figure 3.9. Then we have

$$v_L + v_R = 0$$

and since $v_R = Ri$ and $v_L = L\, \mathrm{d}i/\mathrm{d}t$, then

$$L\frac{\mathrm{d}i}{\mathrm{d}t} + Ri = 0$$

$$\frac{L}{R}\frac{\mathrm{d}i}{\mathrm{d}t} + i = 0 \qquad [29]$$

Example

An inductance of 1.5 H is in series with a resistance of 100 Ω. What will be the differential equation describing how the current in the circuit varies with time if a voltage source of 10 V is switched into the circuit at time $t = 0$?

Equation [27] describes the situation, hence

$$L\frac{\mathrm{d}i}{\mathrm{d}t} + Ri = V$$

$$1.5\frac{\mathrm{d}i}{\mathrm{d}t} + 100i = 10$$

Review problems

8 For the circuit shown in figure 3.8, with the resistance as 10 Ω and the inductance as 0.1 H, what will be the differential equation describing how the current varies with time when the voltage source of 4 V is connected into the circuit at $t = 0$?

9 For the circuit shown in figure 3.9, with the resistance as 50 Ω and the inductance as 0.1 H, what will be the differential equation describing how the current varies with time?

3.4.1 Solving using separation of variables

Consider equation [29],

$$\frac{L}{R}\frac{\mathrm{d}i}{\mathrm{d}t} + i = 0$$

This can have the variables separated to give

$$\frac{\mathrm{d}i}{i} = -\frac{R}{L}\mathrm{d}t$$

Hence

$$\int\frac{\mathrm{d}i}{i} = -\int\frac{R}{L}\mathrm{d}t$$

$$\ln i = -\frac{R}{L}t + A$$

where A is the constant of integration. This equation can be rearranged to give

$$i = e^{-(Rt/L)+A} = e^{A}\,e^{-Rt/L}$$

$$= C\,e^{-Rt/L}$$

where C is a constant. This constant can be determined by taking into account the constraints. Thus, if we have the current as I at $t = 0$ then $C = I$ and so

$$i = I\,e^{-Rt/L} \qquad [30]$$

Equation [28], for the series LR circuit when a step voltage is applied, can also be solved by the separation of variables.

$$\frac{L}{R}\frac{di}{dt} + i = I$$

$$\frac{di}{I - i} = \frac{R}{L}dt$$

$$\int \frac{di}{I - i} = \int \frac{R}{L}dt$$

$$-\ln(I - i) = \frac{R}{L}t + A$$

$$I - i = e^{-(Rt/L)+A} = e^{A}\,e^{-Rt/L}$$

$$i = I - C\,e^{-Rt/L}$$

where C is a constant. The constant can be determined by taking into account the constraints. Thus, if $i = 0$ at $t = 0$ then $C = I$ and so

$$i = I(1 - e^{-Rt/L}) \qquad [31]$$

3.4.2 Assuming a general solution

Consider the differential equation, equation [29], for the decay of current in a series RL circuit.

$$\frac{L}{R}\frac{di}{dt} + i = 0$$

This equation can be written as

$$\frac{di}{dt} = -\frac{R}{L}i$$

It then clearly indicates that the rate of change of the current is proportional to the current and describes an exponential decay (see chapter 2). Thus the solution is of the form

$$i = C\mathrm{e}^{st}$$

where C is a constant. Using this value for the current in the differential equation gives

$$sC\mathrm{e}^{st} = -\frac{R}{L}C\mathrm{e}^{st}$$

Hence $s = -R/L$. The constant C can be determined by the constraints. Thus if $i = I$ at $t = 0$ then $C = I$ and so

$$i = I\mathrm{e}^{-Rt/L} \qquad [32]$$

For the differential equation describing how the current varies with time when there is a voltage source in the circuit, i.e. equation [28],

$$\frac{L}{R}\frac{\mathrm{d}i}{\mathrm{d}t} + i = I$$

then, because this equation is linear (see chapter 1 for a discussion of linear differential equations and chapter 2 for a discussion of the particular integrals and complementary solutions), the solution can be expressed in the form

$$i = i_{\mathrm{n}} + i_{\mathrm{f}}$$

Substituting this in the differential equation enables the differential equation to become

$$\left(\frac{L}{R}\frac{\mathrm{d}i_{\mathrm{n}}}{\mathrm{d}t} + i_{\mathrm{n}}\right) + \left(\frac{L}{R}\frac{\mathrm{d}i_{\mathrm{f}}}{\mathrm{d}t} + i_{\mathrm{f}}\right) = I$$

We thus have the sum of two differential equations, one of which describes a circuit with no voltage source and is considered to be the natural response (giving the complementary solution) and the other which includes the voltage source and is termed the forcing response (giving the particular integral). Thus the natural response is described by

$$\frac{L}{R}\frac{\mathrm{d}i_{\mathrm{n}}}{\mathrm{d}t} + i_{\mathrm{n}} = 0$$

and the forced response by

$$\frac{L}{R}\frac{di_f}{dt} + i_f = I$$

The natural response equation describes an exponential decay (see earlier in this section) and thus has a solution of the form

$$i_n = C\,e^{st}$$

Substituting this value into the differential equation gives

$$\frac{L}{R}sC\,e^{st} + C\,e^{st} = 0$$

Hence $s = -R/L$. Thus

$$i_n = C\,e^{-Rt/L}$$

The differential equation with the voltage source present can be considered to have a solution, i.e. a particular integral (see chapter 2), of the form $i_f = k$. Thus, substituting this into the differential equation gives

$$0 + k = I$$

thus $k = I$. Hence $v_f = I$ and so

$$i = i_n + i_f = C\,e^{-Rt/L} + I$$

The constant C can be determined by the constraints. Thus, if at $t = 0$ we have $i = 0$, then $C = -I$ and so

$$i = I(1 - e^{-Rt/L}) \qquad [33]$$

Example

Determine the differential equation and its solution for the current through the inductor in the circuit shown in figure 3.10.

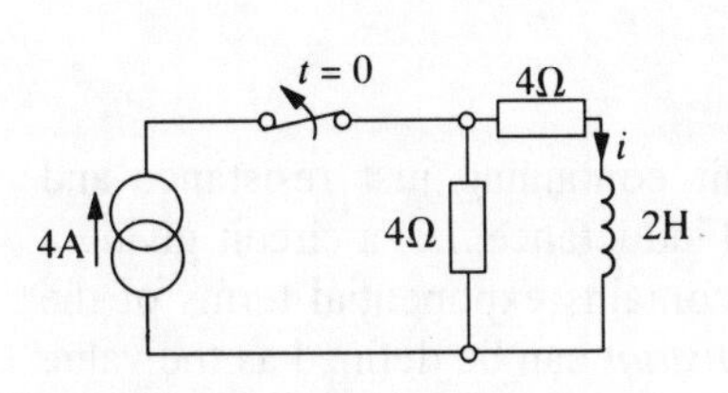

Fig. 3.10 Example

When the switch is opened then

$$v_L + v_R = 0$$

and since $v_R = Ri$ and $v_L = L\,di/dt$, then

$$L\frac{di}{dt} + Ri = 0$$

$$\frac{di}{dt} = -\frac{R}{L}i$$

This equation is of the form characteristic of exponential decay. Thus we can try a solution of the form

$$i = C\,e^{st}$$

Substituting for i in the differential equation then gives

$$sC\,e^{st} = -\frac{R}{L}C\,e^{st}$$

and so $s = -R/L = -8/2$. Thus

$$i = C\,e^{-4t}$$

At $t = 0$ then the current will divide equally between the two branches of the circuit, since they have equal resistances, and so the current through the inductor $i = 2$ A. Thus $C = 2$ A and so

$$i = 2\,e^{-4t}\text{ A}$$

Review problems

10 A coil having a resistance of 40 Ω and an inductance of 0.10 H is connected, at time $t = 0$, across a d.c. supply of 4 V. Derive an equation describing how the current through the coil varies with time.

11 An inductor having a resistance of 2 Ω and an inductance of 2 H is connected in series with a resistance of 4 Ω. Derive an equation describing how the circuit current will vary with time when a voltage of 6 V is applied to the circuit at a time $t = 0$.

12 An inductance L is in series with a resistance R. A voltage source $1t^2$ is applied to the circuit at time $t = 0$. Derive an equation describing how the circuit current varies with time.

3.5 Time constant

The natural response of a circuit containing just resistance and capacitance or just resistance and inductance, i.e. a circuit giving a first-order differential equation, contains exponential terms of the form $e^{-t/RC}$ or $e^{-Rt/L}$. The *time constant* can be defined as the value of the time t which makes the exponential term e^{-1}. Thus for the RC circuit the time constant τ is RC, while for the RL circuit it is L/R.

For a circuit involving a capacitor in series with a resistor the variation with time for the potential difference across the capacitor v_C when a voltage V is connected at time $t = 0$ (equation [13]) is

$$v_C = V(1 - e^{-t/RC})$$

Thus in a time equal to the time constant v_C will rise from 0, the initial value, to $V(1-e^{-1})$ or $0.632V$. In a time equal to 2τ it will rise to $V(1-e^{-2})$ or $0.865V$. In a time equal to 3τ it will rise to $0.950V$, in 4τ to $0.982V$, in 5τ to $0.993V$. Thus it effectively takes a time equal to about five times the time constant to become fully charged.

Example

A circuit consists of an inductance of 1 H in series with a resistance of 10 Ω. At a time $t = 0$ a current source of 2 A is connected to the circuit. What is (a) the time constant of the circuit, (b) the current after a time equal to the time constant?

(a) The time constant is $L/R = 1/10 = 0.1$ s.

(b) The current variation with time is given by equation [33] as

$$i = I(1 - e^{-Rt/L}) = I(1 - e^{-t/\tau})$$

$$= 2(1 - e^{-1}) = 1.26\ \text{V}$$

Review problems

13 A circuit consists of an inductance of 1 H in series with a resistance of 20 Ω. At a time $t = 0$ a current source of 4 A is connected to the circuit. What is (a) the time constant of the circuit, (b) the current after times equal to (i) the time constant, (ii) twice the time constant?

3.5.1 Rate of change at $t = 0$

Consider a circuit involving a capacitor and resistor in series with a voltage V applied to it at time $t = 0$, as in figure 3.2. Then

$$v_R + v_C = V$$

with $i = C\,dv_C/dt$ and $v_R = Ri$. Thus

$$RC\frac{dv_C}{dt} + v_C = V$$

At time $t = 0$ then $v_C = 0$. Thus, the rate of change of the potential difference across the capacitor at time $t = 0$ is

$$\left(\frac{dv_C}{dt}\right)_{t=0} = \frac{V}{RC} = \frac{V}{\tau}$$

The final value of v_C is V. Thus if the initial rate of change of v_C

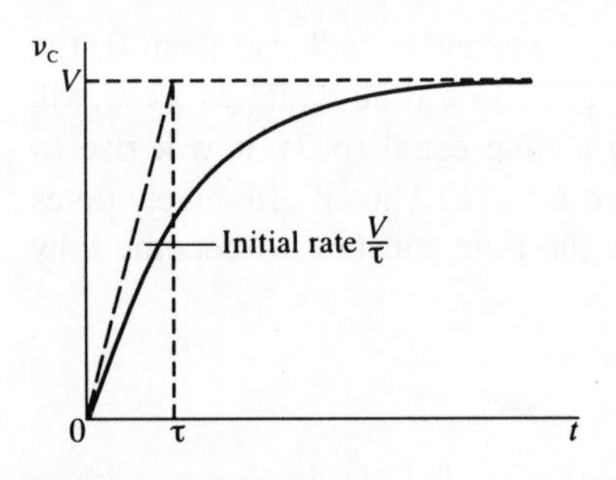

Fig. 3.11 Rise of v_C

was maintained then it would reach the final value in a time equal to the time constant (figure 3.11).

For a series *RL* circuit when a voltage *V* is applied at a time of $t = 0$ then

$$v_L + v_R = V$$

Since $v_R = Ri$ and $v_L = L\,\mathrm{d}i/\mathrm{d}t$ then

$$L\frac{\mathrm{d}i}{\mathrm{d}t} + Ri = V$$

Thus at $t = 0$

$$\left(\frac{\mathrm{d}i}{\mathrm{d}t}\right)_{t=0} = \frac{V}{L}$$

The final current $I = V/R$ and so

$$\left(\frac{\mathrm{d}i}{\mathrm{d}t}\right)_{t=0} = \frac{IR}{L} = \frac{I}{\tau}$$

Thus if the initial rate of change of current was maintained then it would reach the final value in a time equal to the time constant.

Review problems

14 What is the initial rate of change of current in a circuit consisting of an inductance of 1 H in series with a resistance of 10 Ω when current source of 2 A is applied to it?

Further problems

15 Write differential equations describing:
(a) the voltage across a 0.1 μF capacitor when in series with a 2 MΩ resistor and a 3 V step voltage is applied,
(b) the voltage across a charged 1 μF capacitor when it is discharged through a 2 MΩ resistor,
(c) the current through a series circuit of an inductor of inductance 2 H with a 10 Ω resistor when a 4 V step voltage is applied.

16 Determine the differential equation and hence its solution for the potential difference across the capacitor in the circuit shown in figure 3.12.

17 Determine the differential equation and hence its solution for the potential difference across the capacitor in the circuit shown in figure 3.13.

18 Determine the differential equation and hence its solution for the current through the inductor in the circuit shown in figure 3.14.

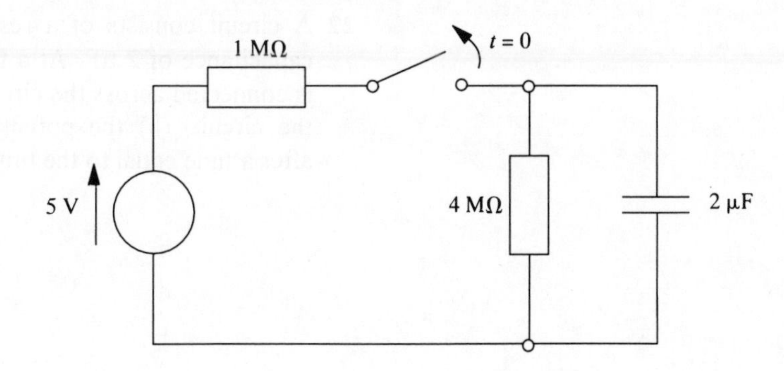

Fig. 3.12 Problem 16

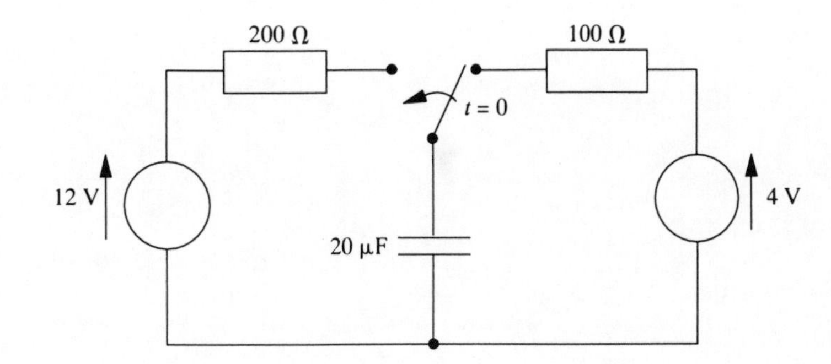

Fig. 3.13 Problem 17

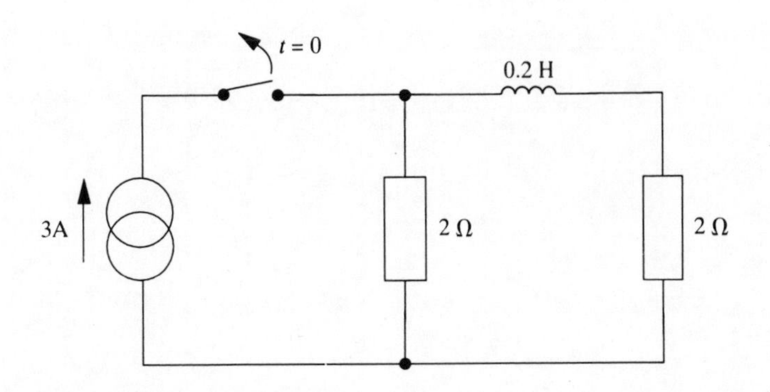

Fig. 3.14 Problem 18

19 A circuit consists of an inductance L in series with a resistance R. A voltage source of $1\mathrm{e}^{-t}$ is applied to the circuit at a time of $t = 0$. Derive the equation relating the circuit current with time.

20 A circuit consists of an inductance of 0.5 H in series with a resistance of 10 Ω. What will be the initial rate of change of current in the circuit when a current source of 4 A is connected to it?

21 A circuit consists of a resistance R in series with an inductance L. When a current source is connected to the circuit, what fraction of the time constant will be required to elapse before the current has risen to half its final value?

22 A circuit consists of a resistance of 1 MΩ in series with a capacitance of 2 μF. At a time $t = 0$ a voltage source of 10 V is connected across the circuit. What is (a) the time constant of the circuit, (b) the potential difference across the capacitor after a time equal to the time constant?

4 First-order differential equations: dynamics

4.1 Dynamics

This chapter is about the application of first-order differential equations to the analysis of the motion resulting from forces being applied to bodies. The following are basic principles involved.

Velocity is defined as the rate of change of displacement with time. *Acceleration* is defined as the rate of change of velocity with time. The *momentum* of a body is defined as the product of its mass m and velocity v, i.e. momentum = mv. When force is in newtons, mass in kilograms and time in seconds then *Newton's second law* can be written as: the rate of change of momentum of a body is equal to the resultant external force acting on it, i.e.

$$F = \frac{\mathrm{d}(mv)}{\mathrm{d}t} \qquad [1]$$

or, if the mass of the body does not change,

$$F = m\frac{\mathrm{d}v}{\mathrm{d}t} \qquad [2]$$

Since the rate of change of velocity is the acceleration, then the acceleration of a constant mass body is proportional to the resultant external force acting on it, i.e.

$$F = ma \qquad [3]$$

When a body is allowed to fall freely in a vacuum it does so with a constant acceleration called the *acceleration due to gravity*. At the surface of the earth this acceleration has a value of about 9.8 m/s^2. The *weight* of a body can be defined as the force needed to prevent it falling and thus has the value, using equation [3], of mg.

4.2 Motion with constant acceleration

Consider an object freely falling from rest, say being allowed to drop vertically downwards from the edge of a cliff. If we assume that the only force acting on the object is that due to gravity, i.e. we are effectively considering the object to be falling in a vacuum and are neglecting any air resistance or buoyancy effects resulting from Archimedes' principle, then the downwards acting force is

$$F = mg$$

But Newton's second law, equation [2], enables us to write this as

$$m\frac{\mathrm{d}v}{\mathrm{d}t} = mg$$

Separation of the variables can be used to obtain a solution to this differential equation, i.e.

$$\int \mathrm{d}v = \int g\,\mathrm{d}t$$

$$v = gt + A$$

where A is the constant of integration. If we take $t = 0$ when the object starts to fall then, since it starts from rest, $v = 0$ at $t = 0$. Thus $A = 0$ and so $v = gt$.

If we wanted to know how the distance x fallen by the object changes with time then, since $v = \mathrm{d}x/\mathrm{d}t$, we can write

$$\frac{\mathrm{d}x}{\mathrm{d}t} = gt$$

Separation of the variables gives

$$\int \mathrm{d}x = \int gt\,\mathrm{d}t$$

$$x = \tfrac{1}{2}gt^2 + A$$

Since $x = 0$ at $t = 0$ then $A = 0$. Thus $x = \frac{1}{2}gt^2$.

The above are very simple problems, the solutions of which could have been obtained by simply using the equations for straight line motion under constant acceleration, i.e. $v = u + at$ and $s = ut + \frac{1}{2}at^2$. However, the method used to tackle the problem can be used in situations where the acceleration is not constant.

Review problems

1 Determine, from the differential equations for the motion, the equations relating (a) velocity and time and (b) distance and time, for an object thrown downwards from the edge of a cliff with an initial velocity u. Assume that the only force acting on the object is gravity.

4.3 Motion with variable acceleration

In the consideration of the motion of falling objects in section 4.1, the only force acting on a falling object was considered to be gravity. This was a simplification since the object is moving in air and there is an air resistance force. The air resistance force F_R is found to be, in general, related to the velocity v of the object by

$$F_R = kv^n \qquad [4]$$

where k is a constant and n has a value $1 \leq n \leq 2$. For low speed motion n is generally taken to have a value of 1, while for high speed motion it is taken to be 2.

Consider an object falling from rest in air when the air resistance force is assumed to be proportional to the velocity. The air resistance force will be in the opposite direction to the gravitational force and thus the resultant force acting on the object is

$$F = mg - kv$$

Thus, using Newton's second law (equation [2]),

$$m\frac{dv}{dt} = mg - kv$$

This differential equation can be solved by separation of the variables, i.e.

$$\int \frac{dv}{mg - kv} = \int \frac{1}{m}\,dt$$

$$\int \frac{dv}{(mg/k) - v} = \int \frac{k}{m}\,dt$$

$$-\ln\left[\left(\frac{mg}{k}\right) - v\right] = \frac{k}{m}t + A$$

$$\frac{mg}{k} - v = e^{-(kt/m)-A} = C\,e^{-kt/m}$$

When $t = 0$ then $v = 0$, thus $C = mg/k$. Hence

$$v = \frac{mg}{k}(1 - e^{-kt/m})$$

Note that as t tends to an infinite value then the velocity tends to the value mg/k. Thus for a falling body with air resistance the velocity does not continue to increase with time but reaches a finite limiting velocity, called the *terminal velocity*. It is this which makes parachute jumping feasible. A parachutist does not keep on increasing in speed as he or she falls but reaches a maximum velocity after a while and then continues with that velocity for the rest of the fall.

The above represents the situation when the force due to air resistance is proportional to the velocity, now consider the case when it is proportional to the square of the velocity. The resultant force acting on the falling object will be

$$F = mg - kv^2$$

Then Newton's second law gives

$$m\frac{dv}{dt} = mg - kv^2$$

This can be written as

$$\frac{1}{v^2 - (mg/k)}\,dv = -\frac{k}{m}\,dt$$

To simplify matters, let $a^2 = mg/k$. Then we have

$$\frac{1}{v^2 - a^2}dv = -\frac{k}{m}\,dt$$

The left-hand side of the equation can be simplified by the use of partial fractions to give

$$\frac{1}{2a}\left(\frac{1}{v-a} - \frac{1}{v+a}\right)dv = -\frac{k}{m}\,dt$$

Integration then gives

$$\frac{1}{2a}[\ln(v-a) - \ln(v+a)] = -\frac{k}{m}t + A$$

$$\ln\left[\frac{v-a}{v+a}\right] = -\frac{2ak}{m}t + 2aA$$

$$\frac{v-a}{v+a} = e^{-2akt/m}\,e^{2aA}$$

Let $C = e^{2aA}$, then

$$v - a = C\,e^{-2akt/m}(v + a)$$

$$v = \frac{a(1 + C\,e^{-2akt/m})}{1 - C\,e^{-2ak/m}}$$

As the time t tends to infinity then v tends to the value a, i.e. $\sqrt{(mg/k)}$. This is then the terminal velocity.

Example

Consider a block of mass 50 kg sliding down an incline of slope 30°, as illustrated in figure 4.1. The block starts from rest on the slope at time $t = 0$. If the coefficient of friction is 0.4 and the force opposing motion due to air resistance is $0.5v$, where v is the velocity, derive an equation stating how the velocity down the slope will vary with time.

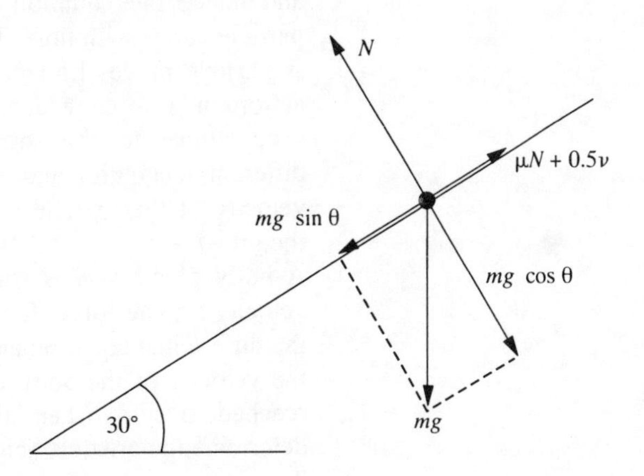

Fig. 4.1 Example

The component of the weight acting down the slope is $mg\sin\theta$ or $50 \times 9.8 \times \sin 30°$, i.e. 245 N. The frictional force opposing the motion down the plane is μN and since the normal force N is equal to $mg\cos\theta$ then the frictional force will be $\mu mg\cos\theta$ and so $0.4 \times 50 \times 9.8 \times \cos 30°$, i.e. 170 N. Also opposing the motion down the plane is the air resistance, a force of $0.5v$. Hence the resultant force acting on the block is

$$F = 245 - 170 - 0.5v = 75 - 0.5v$$

Hence Newton's second law gives

$$50\frac{dv}{dt} = 75 - 0.5v$$

$$\frac{dv}{v-150} = -\frac{1}{100}\,dt$$

Integrating then gives

$$\ln(v-150) = -0.01t + A$$

$$v = e^{A}\,e^{-0.01t} + 150 = Ce^{-0.01t} + 150$$

When $t = 0$ then $v = 0$, hence $C = -150$. Thus

$$v = 150(1 - e^{-0.01t})$$

Review problems

2 A particle moves horizontally along a straight line under the action of a force which results in an acceleration which is proportional to its velocity. Determine the differential equation and hence the equation describing how the velocity of the particle varies with time. The velocity is v_0 at the time $t = 0$.

3 A particle moves horizontally along a straight line under the action of a force which results in an acceleration which is proportional to the square of its velocity. Determine the differential equation and hence the equation describing how the velocity of the particle varies with time. The velocity is v_0 at the time $t = 0$.

4 A body of mass m is shot vertically upwards with an initial velocity u. The force due to air resistance is mkv^2. Determine the differential equation and hence the equation describing how the velocity of the body varies with time. The greatest height reached occurs when the velocity becomes zero, hence determine the greatest height reached.

5 An arrow of mass 0.1 kg is shot vertically upwards with an initial velocity of 90 m/s. The force due to air resistance is $0.004v$. The maximum height is attained when the velocity of the arrow becomes equal to zero. What will be the time taken to reach the maximum height?

6 A motorboat has a mass of 100 kg and has a motor which provides a thrust of 500 N. The force acting on the boat due to water resistance is $10v$, where v is the boat velocity in m/s. Determine the differential equation relating the velocity and time, its solution, and hence the maximum velocity the boat can have.

Further problems

7 A ship of mass 4000 Mg starting from rest is acted on by a constant propeller thrust of 80 000 N. If the water resistance is 10 000v, where v is the velocity in m/s, obtain an equation relating the velocity of the boat with time.

8 An object falls from rest in air and acquires a terminal velocity of 40 m/s. The force due to air resistance is proportional to the square of the velocity. Hence derive an equation relating the velocity with the time of fall.

9 An object of mass m is thrown vertically upwards with an initial velocity u and attains a maximum height of H. If the force due to air resistance is mkv^2 show that

$$H = \frac{1}{2k}\ln\left(1 + \frac{kv^2}{g}\right)$$

and the velocity with which it returns to its starting point is

$$v^2 = \frac{g}{k}(1 - e^{-2kH})$$

10 An object of mass 250 g falls from rest. The force acting on it due to air resistance is $2v$, where v is the velocity in m/s. Derive an equation relating the velocity and time of fall.

11 A particle has an initial velocity of u and suffers a resistive force of $kv^{3/2}$. Derive the differential equation and hence an equation relating the velocity with time.

5 Numerical methods for first-order equations

5.1 Solving first-order differential equations

Ordinary differential equations are often now solved by the use of a computer. The methods used by the computer are not one of the analytical methods discussed in chapter 2 but numerical methods. Indeed, numerical methods are often the only way some differential equations can be solved, whether solved by a computer or not. In this chapter the basic approach to numerical methods is considered, there being many refinements used in practice in order to improve accuracy.

The techniques discussed so far for solving first-order differential equations are termed analytical methods and give rise to solutions in terms of functions such as x, x^2, $\sin x$, e^x, etc. Such solutions are continuous over a range of values of x. Numerical techniques of solving differential equations, however, result in approximate solutions at just a number of discrete values of x. In this chapter the solution of differential equations of the form $dy/dx = f(x, y)$, with some initial value $y = y_0$ at $x = x_0$ being given, are considered. The solutions are obtained for equally spaced values of x, the spacing being called the *step size* and denoted by h. Thus we obtain values of y for x values of 0, h, $2h$, $3h$, etc.

5.2 Euler's method

For a first-order differential equation, dy/dx gives the slope of the tangent at any point on the graph of y against x (see section 1.2). Thus, if we start with the initial value y_0 at $x = 0$ then we can plot this value on a graph of y against x. The differential equation with these values then gives the slope of the tangent at $x = 0$. We now want to find the value of y at $x = h$. If the graph of y against x followed the line of the tangent then we would have (figure 5.1)

$$\text{slope} = \left(\frac{dy}{dx}\right)_{x=0} = \frac{y_1 - y_0}{h}$$

where y_1 is the value of y at $x = h$. Thus

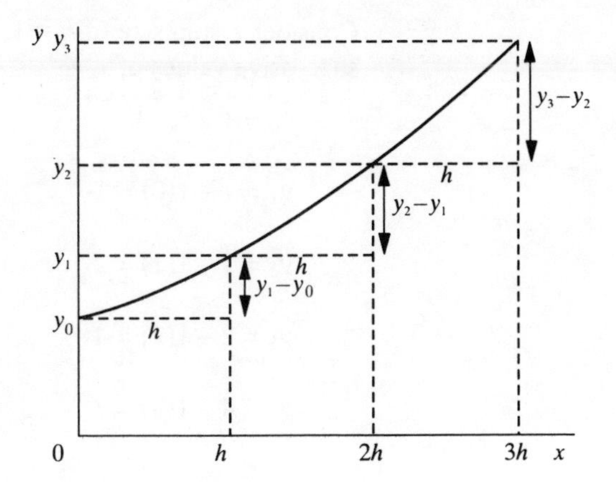

Fig. 5.1 Plotting the graph of y against x

$$y_1 = y_0 + h\left(\frac{dy}{dx}\right)_{x=0}$$

This then enables an estimate of the value of y at $x = h$ to be obtained and hence, using the differential equation, a value of the slope of the graph at this value. We can then use this to obtain a value of y at $x = 2h$.

$$\text{slope} = \left(\frac{dy}{dx}\right)_{x=h} = \frac{y_2 - y_1}{h}$$

$$y_2 = y_1 + h\left(\frac{dy}{dx}\right)_{x=h}$$

We can then systematically, taking it step by step, obtain values of y at $x = 0, h, 2h, 3h$, etc.

The solution graph is built up by repeating the above process over and over again, taking the latest estimated value of y as the starting point each time.

$$y_{n+1} = y_n + h\left(\frac{dy}{dx}\right)_{x=nh} \qquad [1]$$

This step-by-step method of solving a first-order differential equation is known as *Euler's method*.

Example

Use Euler's method to solve the differential equation

$$\frac{dy}{dx} = x$$

if $y = 1$ when $x = 0$.

Consider a step size of $h = 1$. Then using equation [1] we have for successive values of x,

$$y_0 = 1 \qquad x = 0$$

$$y_1 = 1 + 1(0) = 1 \qquad x = 1$$

$$y_2 = 1 + 1(1) = 2 \qquad x = 2$$

$$y_3 = 2 + 1(2) = 4 \qquad x = 3$$

$$y_4 = 4 + 1(3) = 7 \qquad x = 4$$

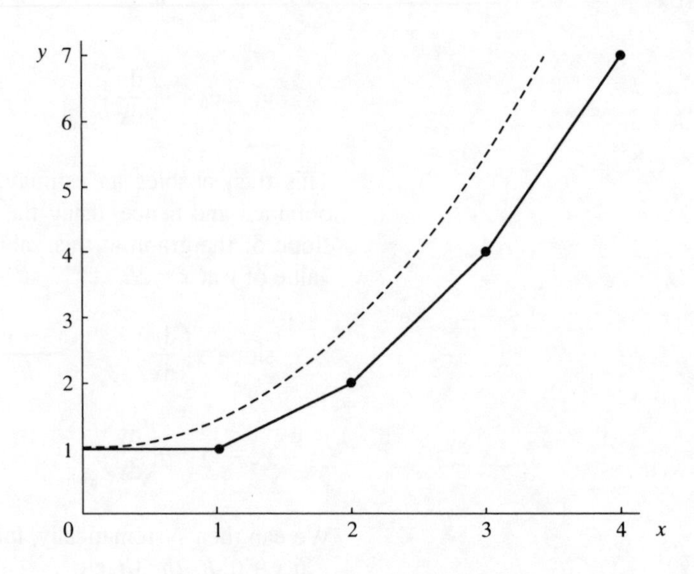

Fig. 5.2 Example

Figure 5.2 shows the solution graph. The analytical solution, which can be obtained by separation of variables, is

$$y = \tfrac{1}{2}x^2 + 1$$

and this is also plotted on the graph in order to show how the numerically obtained solution compares with the analytical one. Notice that as the number of steps increases then so does the error.

Review problems

1 Use Euler's method to solve the following differential equations:

(a) $\dfrac{dy}{dx} = \dfrac{y}{x}$, with $y = 1$ at $x = 1$ and $h = 1$,

(b) $\dfrac{dy}{dx} = \dfrac{y+x}{xy}$, with $y = 2$ at $x = 1$ and $h = 0.1$,

(c) $\dfrac{dy}{dx} = 3x^2 + 1$, with $y = 3$ at $x = 1$ and $h = 0.1$,

(d) $\dfrac{dy}{dx} = 2x + y$, with $y = 1$ at $x = 0$ and $h = 0.2$

2 An object falls with a constant acceleration of 9.8 m/s². Thus $dv/dt = 9.8$ m/s. If the object starts at time $t = 0$ with an initial downward velocity of 2 m/s, use Euler's method with $h = 0.1$ s to obtain the data for a graph of velocity against time.

3 The rate at which the potential difference across a capacitor decreases with time when it is allowed to discharge through a resistor is given by

$$\frac{dv_C}{dt} = -1v_C$$

Use Euler's method with $h = 0.1$ to solve the differential equation if $v_C = 2$ at $t = 0$.

5.2.1 Error estimates for Euler's method

At any one value of x the error due to Euler's method is the difference between the actual value, obtained by analytic means, and the value given by Euler's method. This is illustrated in figure 5.3. The size of the error will depend on the size of the step h; the smaller the step the smaller will be the error.

Consider the solution of the differential equation

$$\frac{dy}{dx} = x + y, \text{ with } y = 1 \text{ at } x = 0$$

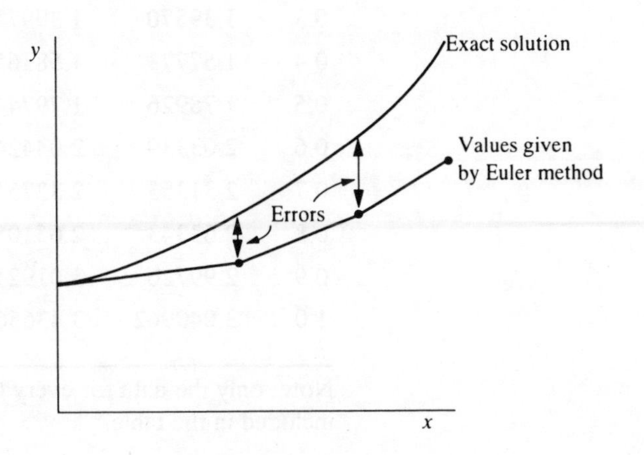

Fig. 5.3 Errors

The analytical solution is $y = 2\,e^x - x - 1$ (a similar equation was solved in section 2.3 by the use of an integrating factor). Suppose we require the numerical solution over the range $x = 0$ to $x = 1$. We can divide this interval into ten steps and so have a step size of $h = 0.1$. Alternatively we could divide the interval into one hundred steps and so have a step size of $h = 0.01$. Tables 5.1(a) and (b) show the results. The smaller value of h gives a better

Table 5.1(a) $h = 0.1$

x	Computed y	True y	Error
0	1.00000	1.00000	0.00000
0.1	1.10000	1.11034	0.00034
0.2	1.22000	1.24281	0.00281
0.3	1.36200	1.39972	0.03792
0.4	1.52820	1.58365	0.05545
0.5	1.72102	1.79744	0.07642
0.6	1.94312	2.04424	0.10112
0.7	2.19743	2.32751	0.13008
0.8	2.48718	2.65108	0.16390
0.9	2.81590	3.01921	0.20331
1.0	3.18749	3.43656	0.24907

Table 5.1(b) $h = 0.01$

x	Computed y	True y	Error
0	1.00000	1.00000	0.00000
0.1	1.10924	1.11034	0.00110
0.2	1.24038	1.24281	0.00243
0.3	1.39570	1.39972	0.00402
0.4	1.57773	1.58365	0.00592
0.5	1.78926	1.79744	0.00818
0.6	2.03339	2.04424	0.01085
0.7	2.31353	2.32751	0.01398
0.8	2.63343	2.65108	0.01765
0.9	2.99726	3.01921	0.02195
1.0	3.040962	3.43656	0.02694

Note: only the data for every tenth step have been included in the table.

approximation to the true values. However, it should also be noted that the errors increase, with both values of h, as x increases.

5.3 The improved Euler method

Euler's method uses the slope at some point x to predict the value of y at $x + h$ (figure 5.4). It is thus assuming that the slope at x is the slope of the solution over the interval $x + h$. We then use this assumption to deduce a value for y at $x + h$ (as in equation [1]).

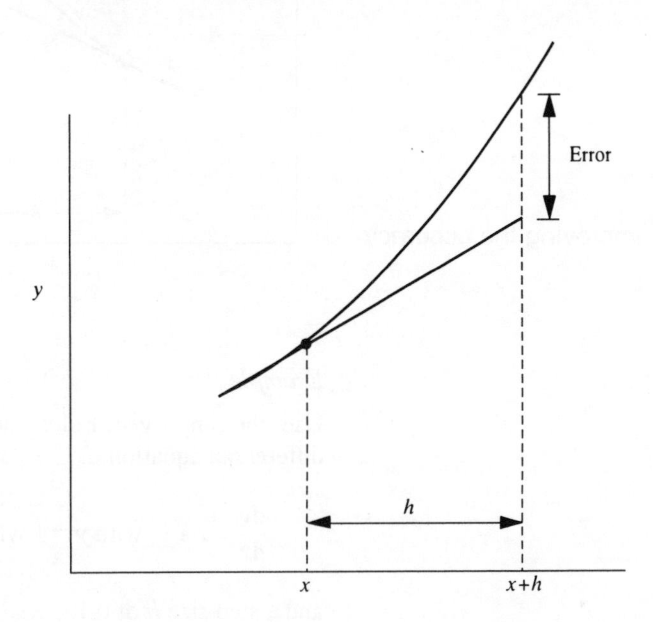

Fig. 5.4 Error

The error with the Euler method arises because we assume that the slope at x is the slope over the interval x to $x + h$. Suppose we now just regard the value of y predicted by this for $x + h$ as a first estimate. We can then use this estimate to obtain a value for the slope at $x + h$. It seems reasonable to assume that a better approximation to the slope over the interval x to $x + h$ will be the average of the slopes at x and $x + h$. Thus the average is taken of the slope at x and that predicted for $x + h$ and used as the slope over the interval x to $x + h$ to obtain an improved estimate of y at $x + h$. Figure 5.5 illustrates this.

The second estimate of y at $x + h$ can then be used to obtain a second estimate of the slope at $x + h$ and so a new average obtained and hence a third estimate of y obtained. This process can be repeated for yet more improved estimates, the result at each estimate getting nearer and nearer to the true value.

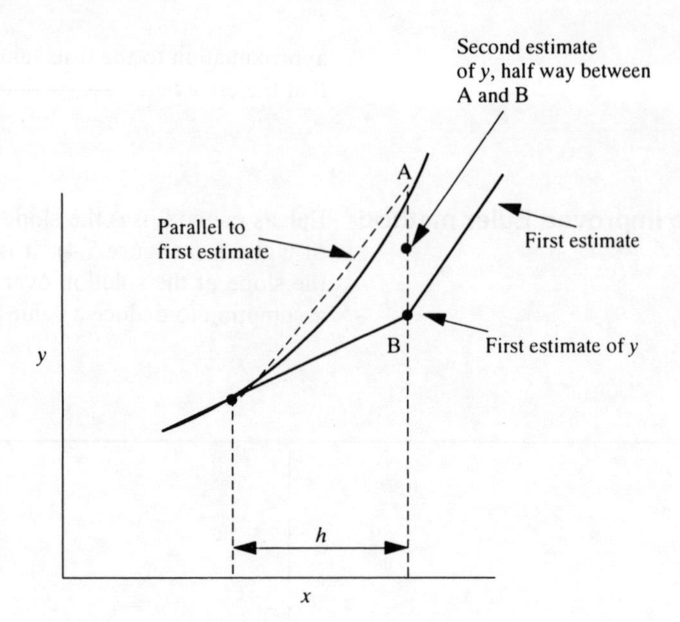

Fig. 5.5 Improving the accuracy

Example

Use the improved Euler method to obtain a solution to the differential equation

$$\frac{dy}{dx} = x + y, \text{ with } y = 1 \text{ when } x = 0$$

and a step size h of 0.1.

At $x = 0$ we have $y = 1$. Thus, using equation [1],

$$y_{n+1} = y_n + h\left(\frac{dy}{dx}\right)_{x=nh}$$

at $x = 1h = 0.1$ the estimated value of y_{n+1} is

$$y_{n+1} = 1 + 0.1(1) = 1.1$$

Using this value we can obtain the slope at y_{n+1} as $0.1 + 1.1 = 1.2$. The average of this slope with that used in the first estimate of 1 is $\frac{1}{2}(1 + 1.2) = 1.1$. Thus, using this as the revised slope

$$y_{n+1} = 1 + 0.1(1.1) = 1.11$$

For the second step, at $x = 2h = 0.2$ the estimated value of y_{n+2} is

$$y_{n+2} = 1.11 + 0.1(1.21) = 1.231$$

Using this value we can obtain the slope at y_{n+2} as 0.2 + 1.231 = 1.431. The average of this slope with that used in the first estimate of 1.1 is $\frac{1}{2}(1.21 + 1.431) = 1.3205$. Thus, using this as the revised slope

$$y_{n+2} = 1.11 + 0.1(1.3205) = 1.2421$$

Table 5.2 shows the results of continuing this calculation for

Table 5.2 Euler and improved Euler data for $h = 0.1$

x	Unimproved value for y	Improved value for y	True y
0	1.0000	1.0000	1.0000
0.1	1.1000	1.1000	1.1103
0.2	1.2200	1.2421	1.2428
0.3	1.3620	1.3985	1.3997
0.4	1.5282	1.5818	1.5837
0.5	1.7210	1.7949	1.7974
0.6	1.9431	2.0409	2.0442
0.7	2.1974	2.3231	2.3275
0.8	2.4872	2.6456	2.6511
0.9	2.8159	3.0124	3.0192
1.0	3.1875	3.4282	3.4366

further steps and compares the result with the unimproved Euler method and the analytical result (these were calculated earlier for table 5.1).

Review problems

4 Determine the solution of the following differential equation using the Euler and the improved Euler method with a step size of $h = 0.1$ for the first four steps.

$$\frac{dy}{dx} = x, \text{ with } y = 1 \text{ when } x = 0$$

5 Determine the solution of the following differential equation using the Euler and the improved Euler method with a step size of $h = 0.2$ for the first four steps.

$$\frac{dy}{dx} = 2x + y, \text{ with } y = 1 \text{ when } x = 0$$

5.4 Taylor series method

Another way that can be used to improve the Euler method is to use a power series. A graph of y against x for some function can be approximated by the sum of a number of positive integral powers of x. Thus we can have

$$y = f(x) = a_0 + a_1 x + a_2 x^2 + a_3 x^3 + \ldots \qquad [2]$$

where a_0, a_1, etc., are constants. Such a series can be differentiated term by term to give

$$\frac{dy}{dx} = a_1 + 2a_2 x + 3a_3 x^2 + 4a_4 x^3 + \ldots$$

$$\frac{d^2y}{dx^2} = 2a_2 + 3 \times 2a_3 x + 4 \times 3a_4 x^2 + \ldots$$

$$\frac{d^3y}{dx^3} = 3 \times 2a_3 + 4 \times 3 \times 2a_4 x + \ldots$$

and so on. If we now consider the values of the above when $x = 0$, then the above expressions give

$$(y)_{x=0} = a_0$$

$$\left(\frac{dy}{dx}\right)_{x=0} = a_1$$

$$\left(\frac{d^2y}{dx^2}\right)_{x=0} = 2a_2$$

$$\left(\frac{d^3y}{dx^3}\right)_{x=0} = 2 \times 3a_3$$

Thus, substituting these values for the coefficients in the series, gives

$$y = (y)_{x=0} + x\left(\frac{dy}{dx}\right)_{x=0} + \frac{x^2}{2!}\left(\frac{d^2y}{dx^2}\right)_{x=0} + \frac{x^3}{3!}\left(\frac{d^3y}{dx^3}\right)_{x=0} + \ldots \qquad [3]$$

This series is known as the *Maclaurin series*. This series tells us the value of y at any point on the curve in terms of the zero point conditions.

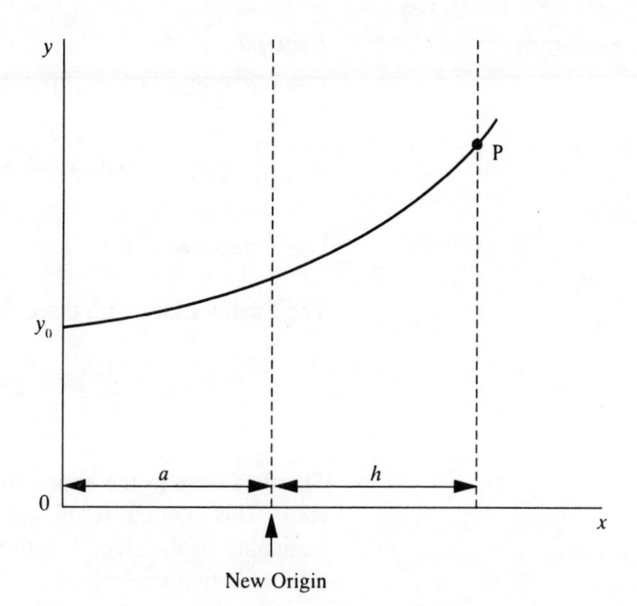

Fig. 5.6 Change of origin

Suppose we now want the value of y at some point P in terms of the values at some other point than $x = 0$, say at $x = h$ (figure 5.6). We can do this by considering shifting the origin of the graph to the value of x concerned. The values of x referred to this new origin are $(x - a)$. If we let $h = x - a$, then equation [3] becomes

$$y = (y)_{x=a} + h\left(\frac{\mathrm{d}y}{\mathrm{d}x}\right)_{x=a} + \frac{h^2}{2!}\left(\frac{\mathrm{d}^2y}{\mathrm{d}x^2}\right)_{x=a} + \frac{h^3}{3!}\left(\frac{\mathrm{d}^3y}{\mathrm{d}x^3}\right)_{x=a} + \ldots \quad [4]$$

This is known as the *Taylor series*.

In terms of the symbols used earlier in this chapter for the Euler method, for one step the Taylor series can be written as

$$y_{n+1} = y_n + h\left(\frac{\mathrm{d}y}{\mathrm{d}x}\right)_{x=nh} + \frac{h^2}{2!}\left(\frac{\mathrm{d}^2y}{\mathrm{d}x^2}\right)_{x=nh} + \frac{h^3}{3!}\left(\frac{\mathrm{d}^3y}{\mathrm{d}x^3}\right)_{x=nh} + \ldots \quad [5]$$

The first two terms in the series give the Euler equation [1] developed earlier. However, provided the Taylor series converges, using more terms in the series at each step improves the accuracy.

Example

Use the Taylor series with three terms to find the solution of

$$\frac{dy}{dx} = y^2 - x, \text{ with } y = 1 \text{ when } x = 0$$

Use a step size of 0.1.

The Taylor series with three terms is

$$y_{n+1} = y_n + h\left(\frac{dy}{dx}\right)_{x=nh} + \frac{h^2}{2!}\left(\frac{d^2y}{dx^2}\right)_{x=nh}$$

The first term in the expression will be the initial value of y for a step. The second term will be the value of $h(y^2 - x)$ at the beginning of the step. The third term will be the value obtained by differentiating $y^2 - x$ by x.

$$\frac{h^2}{2}\frac{d^2y}{dx^2} = \frac{h^2}{2}\left(2y\frac{dy}{dx} - 1\right)$$

$$= \frac{h^2}{2}[2y(y^2 - x) - 1]$$

For $x = 0$ we have $y = 1$. Then for $x = 0.1$, i.e. after one step, we have

$$y = 1 + 0.1(1^2 - 0) + \frac{0.1^2}{2}[2(1^2 - 0) - 1]$$

$$= 1.105$$

For $x = 0.2$, i.e. two steps, we have

$$y = 1.105 + 0.1(1.105^2 - 0.1)$$

$$+ \frac{0.1^2}{2}[2 \times 1.105(1.105^2 - 0.1) - 1]$$

$$= 1.2245$$

For $x = 0.3$, i.e. three steps, we have

$$y = 1.2245 + 0.1(1.2245^2 - 0.2)$$

$$+ \frac{0.1^2}{2}[2 \times 1.2245(1.2245^2 - 0.2) - 1]$$

$$= 1.3654$$

Review problems

6 Use (a) three terms, (b) five terms of the Taylor series to determine the value of y after a step of $x = 0.2$ for the following differential equation:

$$\frac{dy}{dx} = x - y, \text{ with } y = 1 \text{ when } x = 0$$

5.5 The Runge-Kutta method

Since it requires the evaluation of derivatives, determining the terms in the Taylor series can be a rather complicated and tedious process. A method which overcomes this is the Runge-Kutta method. This just requires four simple evaluations of the function being solved and gives a result which is equivalent to using the Taylor series up to the fourth term.

Consider a differential equation which is some function of x and y, i.e. $dy/dx = f(x,y)$, and which has an initial value of y_0 at x_0. We can use the Euler method to give a first estimate of the value y_1 after one step in x of h. Figure 5.7 shows the result, the line AB having been drawn to represent the slope of the graph at y_0. The slope of AB we designate as k_1/h, with k_1 being the change in y occurring in that step from A to B. Hence k_1/h is the value of the differential equation with the values of x_0 and y_0 used, i.e.

$$\frac{k_1}{h} = f(x_0, y_0) \qquad [6]$$

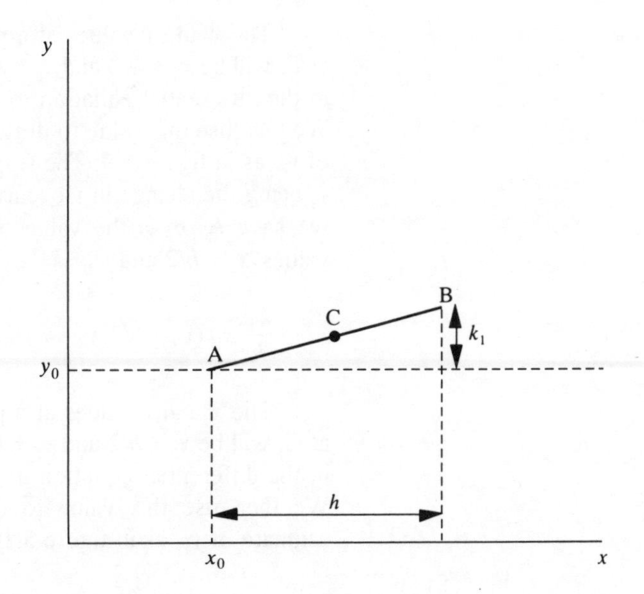

Fig. 5.7 Determining k_1

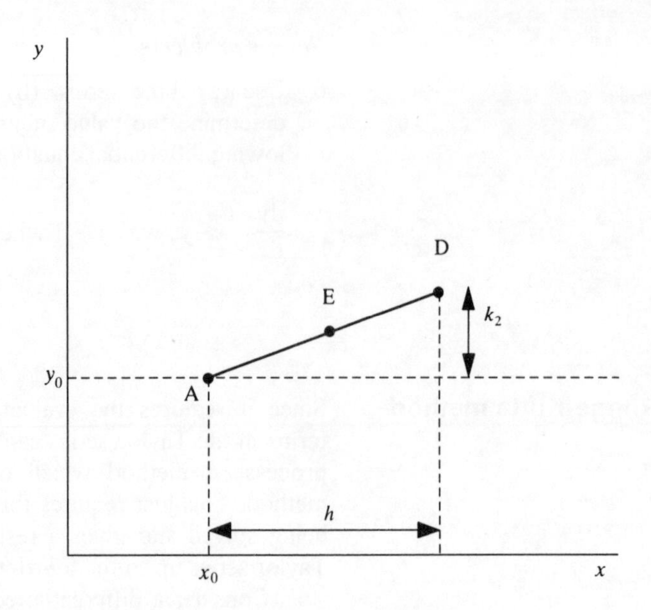

Fig. 5.8 Determining k_2

The x and y values at a point halfway along the line AB, i.e. at C, will be $x_0 + h/2$ and $y_0 + k_1/2$. We now substitute these values in the differential equation to arrive at a new value for the slope. We then use this value to draw a line AD to give a revised estimate of y_1, as in figure 5.8. The slope of AD we designate as k_2/h with k_2 being the change in y that occurs in that step from A to D. Hence k_2/h is the value of the differential equation with the values $x_0 + h/2$ and $y_0 + k_1/2$ used, i.e.

$$\frac{k_2}{h} = f(x_0 + h/2, y_0 + k_1/2) \tag{7}$$

The x and y values at a point halfway along the line AD, i.e. at E, will be $x_0 + h/2$ and $y_0 + k_2/2$. We now substitute these values in the differential equation to arrive at a new value for the slope. We then use this value to draw a line AF to give a revised estimate of y_1, as in figure 5.9. The slope of AF we designate as k_3/h, with k_3 being the change in y occurring in that step from A to E. Hence we have k_3/h as the value of the differential equation with the values $x_0 + h/2$ and $y_0 + k_2/2$ used, i.e.

$$\frac{k_3}{h} = f(x_0 + h/2, y_0 + k_2/2) \tag{8}$$

The x and y values at a point halfway along the line AF, i.e. at G, will be $x_0 + h/2$ and $y_0 + k_3/2$. We now substitute these values in the differential equation to arrive at a new value for the slope. We then use this value to draw a line AH to give a revised estimate of y_1, as in figure 5.10. The slope of AH we designate as

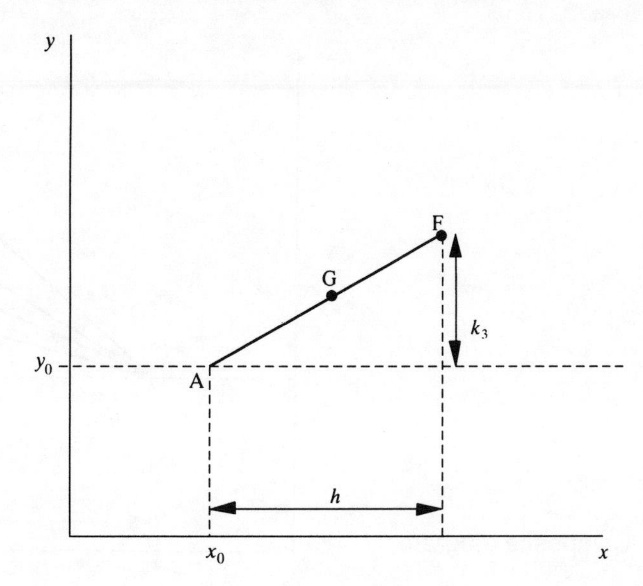

Fig. 5.9 Determining k_3

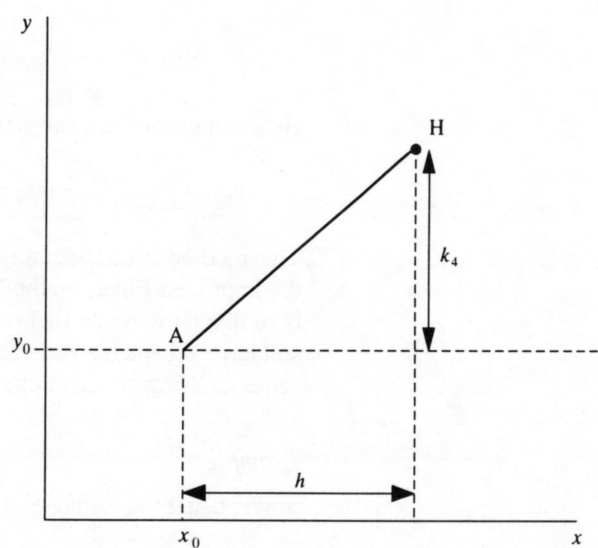

Fig. 5.10 Determining k_4

k_4/h, with k_4 being the change in y occurring in that step from A to E. Hence k_4/h is the value of the differential equation with the values $x_0 + h/2$ and $y_0 + k_3/2$ used, i.e.

$$\frac{k_4}{h} = f(x_0 + h/2, y_0 + k_4/2) \qquad [9]$$

Figure 5.11 shows the composite diagrams with the slopes obtained at the various stages above all being shown on the same figure. A weighted average value of k is then used to determine the corrected estimate of y_1, with

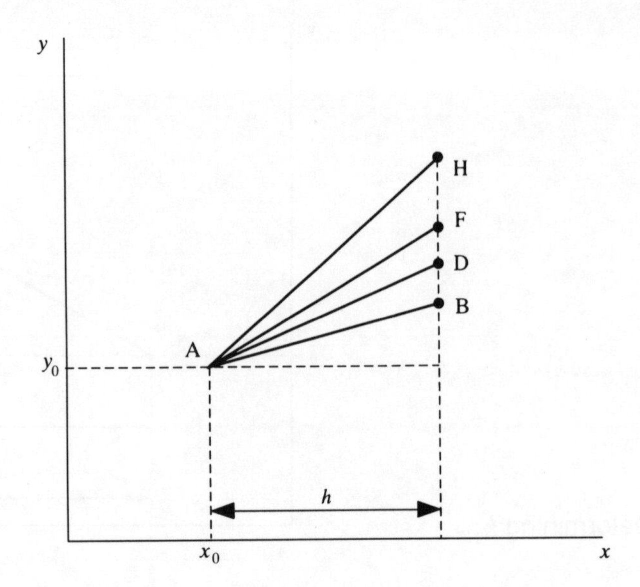

Fig. 5.11 Composite diagram

$$k = \tfrac{1}{6}(k_1 + 2k_2 + 2k_3 + k_4) \qquad [10]$$

Hence in general we can write for a step

$$y_{n+1} = y_n + \tfrac{1}{6}(k_1 + 2k_2 + 2k_3 + k_4) \qquad [11]$$

This method is considerably more accurate than the Euler, or even the improved Euler, method and is very widely used. The accuracy is so much improved that it is possible to take larger step sizes and still have acceptable accuracy. It has the advantage over the Taylor series of being easier to use.

Example

Solve, using the Runge-Kutta method, the differential equation

$$\frac{\mathrm{d}y}{\mathrm{d}x} = x + y, \text{ with } y = 1 \text{ when } x = 0$$

Take the step size h to be 0.1.

Using equation [6],

$$k_1 = hf(x_0, y_0) = 0.1(0 + 1) = 0.1$$

Using equation [7]

$$k_2 = hf(x_0 + h/2, y_0 + k_1/2)$$

$$= 0.1(0 + 0.05 + 1 + 0.05) = 0.11$$

Using equation [8]

$$k_3 = hf(x_0 + h/2, y_0 + k_2/2)$$

$$= 0.1(0 + 0.05 + 1 + 0.055) = 0.1105$$

Using equation [9]

$$k_4 = hf(x_0 + h/2, y_0 + k_4/2)$$

$$= 0.1(0 + 0.05 + 1 + 0.05525) = 0.12105$$

Thus the weighted average of the k values gives (equation [10])

$$y_1 = y_0 + \tfrac{1}{6}(k_1 + 2k_2 + 2k_3 + k_4)$$

$$= 1 + \tfrac{1}{6}(0.1 + 2 \times 0.11 + 2 \times 0.1105 + 0.12105)$$

$$= 1.11034$$

This procedure can then be repeated for each of the steps. Table 5.3 gives the results.

Table 5.3 Runge-Kutta data

x	Runge-Kutta value for y	True value for y
0	1.00000	1.00000
0.1	1.11034	1.11034
0.2	1.24281	1.24281
0.3	1.39972	1.39972
0.4	1.58365	1.53865
0.5	1.79744	1.79744
0.6	2.04424	2.04424
0.7	2.32750	2.32750
0.8	2.65108	2.65108
0.9	3.01920	3.01920
1.0	3.43656	3.43656

Review problems

7 Use the Runge-Kutta method to solve the differential equation

$$\frac{dy}{dx} = y + 2x, \text{ with } y = 1 \text{ at } x = 0$$

Take the step size to be 0.2.

8 Use the Runge-Kutta method to estimate the value of y at x = 0.5 for

$$\frac{dy}{dx} = x^2 + y^2, \text{ with } y = 1 \text{ at } x = 0$$

Take the step size to be 0.1.

9 Use the Runge-Kutta method to estimate the value of y at x = 0.5 for

$$\frac{dy}{dx} = y - x, \text{ with } y = 2 \text{ at } x = 0$$

Take the step size to be 0.1.

Further problems

10 Use Euler's method to determine the solution of the following differential equations:

(a) $\frac{dy}{dx} = x + y$, when $y = 0$ at $x = 0$ with $h = 0.1$,

(b) $\frac{dy}{dx} = x + y$, when $y = 0$ at $x = 0$ with $h = 0.2$,

(c) $\frac{dy}{dx} = 2xy + 1$, when $y = 0$ at $x = 0$ with $h = 0.1$,

(d) $\frac{dy}{dx} = 2 - x$, when $y = 1$ at $x = 0$ with $h = 0.1$

11 An electrical circuit consists of an inductance of 4 H in series with a resistance of 2 Ω. When a constant voltage of 20 V is connected to the circuit at time t = 0 then the differential equation describing how the circuit current varies with time is

$$2\frac{di}{dt} + i = 10$$

Use Euler's method with h = 0.2 s to determine the data for a graph of current against time.

12 Determine the solution of the following differential equation using the Euler and the improved Euler methods with a step size of 0.1 for the first four steps.

$$\frac{dy}{dx} = 2x, \text{ with } y = 1 \text{ when } x = 0$$

13 Determine, using the improved Euler method, the value of y at $x = 0.5$ for the differential equation

$$\frac{dy}{dx} = y + 1, \text{ with } y = 1 \text{ when } x = 0$$

Use a step size of $h = 0.1$.

14 Determine, using the improved Euler method, the value of y at $x = 1$ for the differential equation

$$\frac{dy}{dx} = x - 2y, \text{ with } y = 1 \text{ when } x = 0$$

Use a step size of 0.2.

15 Use (a) three terms, (b) five terms of the Taylor series to determine the value of y after one step of $x = 0.1$ for the differential equation

$$\frac{dy}{dx} = y^2 + 3x, \text{ with } y = 1 \text{ when } x = 0$$

16 Use (a) three terms, (b) five terms of the Taylor series to determine the value of y after one step of $x = 0.1$ for the differential equation

$$\frac{dy}{dx} = x - y^2, \text{ with } y = 1 \text{ when } x = 0$$

17 Use the Runge-Kutta method to determine the value of y at $x = 0.2$ for

$$\frac{dy}{dx} = \frac{y^2}{x+1}, \text{ with } y = 1 \text{ when } x = 0$$

Use a step size of 0.05.

18 Use the Runge-Kutta method to determine the value of y at $x = 0.4$ for

$$\frac{dy}{dx} = x^2 + y^2, \text{ with } y = 0 \text{ when } x = 0$$

Use a step size of 0.1.

19 Use the Runge-Kutta method to determine the value of y at $x = 1.0$ for

$$\frac{dy}{dx} = x - 2y, \text{ with } y = 1 \text{ when } x = 0$$

Use a step size of 0.2.

20 Use the Runge-Kutta method to determine the value of y at $x = 0.4$ for

$$\frac{dy}{dx} = y - 2, \text{ with } y = 1 \text{ at } x = 0$$

Use a step size of 0.1.

21 Use the Runge-Kutta method to determine the solution of the following differential equation from $x = 0$ to 1.0.

$$\frac{dy}{dx} = x + y, \text{ with } y = 0 \text{ at } x = 0$$

Use a step size of 0.2.

22 A stone falls from rest with an acceleration due to gravity of 10 m/s^2 (a rounded figure for the acceleration due to gravity) and experiences air resistance which results in a retardation of $0.1v$ m/s^2. Hence

$$\frac{dv}{dt} = 10 - 0.1v$$

Determine by means of a numerical method the velocity after (a) 1 s, (b) 4 s. You might like to try the Runge-Kutta method with a step size of 0.2.

23 An object of mass m falls from rest in a viscous medium. The upthrust acting on it is $\frac{1}{4}mg$ and the viscous drag is mv. Hence

$$m\frac{dv}{dt} = mg - \tfrac{1}{4}mg - mv$$

Use a numerical method to determine the velocity after 1 s.

24 A ball is thrown upwards with an initial velocity of 50 m/s. The force due to air resistance is $0.1mv$. Thus, taking the acceleration due to gravity as 10 m/s^2,

$$m\frac{dv}{dt} = -10m - 0.1mv$$

Use a numerical method to determine the velocity after (a) 1 s, (b) 2 s.

25 An electrical circuit consists of an uncharged capacitor in series with a resistor. At the time $t = 0$ if a constant voltage of V is applied to the circuit

$$RC\frac{dv_C}{dt} + v_C = V$$

If the resistance is 2 MΩ, the capacitance 1 μF and V is 5 V, use a numerical method to determine the voltage v_C across the capacitor after 1 s.

6 Higher order differential equations

6.1 Higher order differential equations

A second-order differential equation is one that contains a second derivative term but no higher derivative. Second-order linear differential equations with constant coefficients are very commonly encountered in engineering and science.

For example, since acceleration is rate of change of velocity v with time t and velocity is rate of change of displacement y with time, acceleration can be written as the rate of change of the rate of change of displacement with time, i.e.

$$\text{acceleration} = \frac{dv}{dt} = \frac{d}{dt}\left(\frac{dy}{dt}\right) = \frac{d^2y}{dt^2}$$

Thus for a freely falling object when the acceleration is just the acceleration due to gravity g, then the differential equation describing how the displacement varies with time is

$$\frac{d^2y}{dt^2} = g \qquad [1]$$

Another example is that of the displacement of a mass m freely oscillating on the end of a spring, or indeed any form of mechanical oscillatory system, when there is some form of damping. Such a system will have a differential equation of the form

$$m\frac{d^2y}{dt^2} + c\frac{dy}{dt} + ky = 0$$

Such oscillatory systems are discussed in more detail in chapter 7.

An electrical example of a second-order differential equation is given by an electrical circuit containing resistance, capacitance and inductance in series. The differential equation describing how the potential difference v_C across the capacitor varies with time when a voltage V is applied at time $t = 0$ is

$$LC\frac{d^2v_C}{dt^2} + RC\frac{dv_C}{dt} + v_C = V$$

See chapter 8 for a discussion of such equations.

A third-order differential equation contains a third derivative term but no higher derivative, a fourth-order differential equation a fourth derivative term but not higher. This chapter is primarily about the solution of second-order differential equations with constant coefficients, since they are so common in engineering and science, with just brief indications of how higher order equations can be solved.

6.2 Arbitrary constants

Consider the differential equation given earlier (equation [1]) for a freely falling object. If we take the acceleration due to gravity to be 10 m/s^2 then it becomes

$$\frac{d^2y}{dt^2} = 10 \qquad [2]$$

If we integrate both sides of the equation with respect to dt, then

$$\int \frac{d^2y}{dt^2}\,dt = \int 10\,dt$$

$$\frac{dy}{dt} = 10t + A \qquad [3]$$

where A is the constant of integration. If we again integrate with respect to dt, then

$$\int \frac{dy}{dt}\,dt = \int (10t + A)\,dt$$

$$y = 5t^2 + At + B \qquad [4]$$

where B is a constant of integration. Thus, the general solution of the second-order differential equation contains two arbitrary constants. This is true of all second-order differential equations. The number of arbitrary constants in the general solution of a differential equation is equal to the order of that equation. Thus, for example, a fourth-order differential equation will have four arbitrary constants.

With the second-order differential equation and the two arbitrary constants in its solution, it will require two initial conditions to determine them. Thus, for the freely falling object we might have an initial condition for the displacement y and the velocity v, i.e. $\mathrm{d}y/\mathrm{d}t$.

$$y = 0 \text{ at } t = 0 \text{ and } \frac{\mathrm{d}y}{\mathrm{d}t} = 0 \text{ at } t = 0$$

Thus, using the initial condition $\mathrm{d}y/\mathrm{d}t = 0$ when $t = 0$ and equation [3]

$$0 = 10 \times 0 + A$$

Hence $A = 0$. Using the condition $y = 0$ at $t = 0$ with equation [4] then

$$0 = 5 \times 0 + 0 + B$$

and so $B = 0$. Thus the solution with these initial conditions is

$$y = 5t^2$$

Thus to determine the particular solution of a second-order differential equation, two initial conditions are required. To determine the particular solution of a third-order differential equation, three initial conditions are required, for a fourth-order equation four initial conditions.

The above technique of integrating both sides of a second-order differential equation with respect to x applies when the second-order equation is of the form

$$\frac{\mathrm{d}^2y}{\mathrm{d}x^2} = f(x)$$

Two direct integrations give the general solution.

Example

Solve the following differential equation

$$\frac{\mathrm{d}^2y}{\mathrm{d}x^2} = 3, \text{ with } y = 2 \text{ and } \frac{\mathrm{d}y}{\mathrm{d}x} = 4 \text{ at } x = 0$$

Integrating both sides of the differential equations with respect to x gives

$$\int \frac{\mathrm{d}^2y}{\mathrm{d}x^2}\,\mathrm{d}x = \int 3\,\mathrm{d}x$$

$$\frac{dy}{dx} = 3x + A$$

Since $dy/dx = 4$ when $x = 0$, then $A = 4$. Integrating again with respect to x gives

$$y = \tfrac{3}{2}x^2 + Ax + B$$

Since $y = 2$ when $x = 0$, then $B = 2$. Thus the particular solution is

$$y = \tfrac{3}{2}x^2 + 4x + 2$$

Review problems

1 Solve the following differential equations by two stages of integration.

(a) $\dfrac{d^2y}{dx^2} = 2$, with $y = 2$ and $\dfrac{dy}{dx} = 3$ at $x = 0$,

(b) $\dfrac{d^2y}{dx^2} = 6$, with $y = 2$ at $x = 0$ and $\dfrac{dy}{dx} = 6$ at $x = 1$,

(c) $\dfrac{d^2y}{dx^2} = 4x$, with $y = 1$ and $\dfrac{dy}{dx} = 2$ at $x = 0$,

(d) $\dfrac{d^2y}{dx^2} = 6x^2 + 2$, with $y = 0$ and $\dfrac{dy}{dx} = 0$ at $x = 0$

2 The deflection y at a distance x from the fixed end of a cantilever is given by

$$EI\frac{d^2y}{dx^2} = FL - Fx$$

where F is the load at the free end and L the length of the cantilever. Solve this differential equation given that $y = 0$ and $dy/dx = 0$ at $x = 0$.

6.3 The homogeneous linear second-order equation

A homogeneous second-order differential equation with constant coefficients a, b and c is of the basic form

$$a\frac{d^2y}{dx^2} + b\frac{dy}{dx} + cy = 0 \qquad [5]$$

If the right-hand side of the equation was not zero then the equation would be termed non-homogeneous (see section 6.4). It is zero because there is no externally applied input to the system. Thus if, for example, the equation describes the oscillations of a weight suspended from a spring then it is just the natural oscillations of the system when there is no externally applied force, such as the support to which the spring is attached being vibrated by some externally applied force. The equation could describe the current in a circuit containing resistance, capacitance and inductance when there is no voltage input to the circuit. Because it is zero the differential equation is often described as being the natural or free motion equation. Such second-order linear differential equations with constant coefficients occur often in engineering and science. Chapters 8 and 9 show their applications in the description of mechanical and electrical oscillations.

Since the solution of the first-order linear homogeneous differential equation with constant coefficients, namely

$$\frac{dy}{dx} + ky = 0$$

is of the general form $y = C\,e^{kx}$ it seems reasonable to consider that

$$y = A\,e^{sx} \qquad [6]$$

with A and s being constants, might be the form of the solution of the second-order differential equation. Thus, since $dy/dx = As\,e^{sx}$ and $d^2y/dx^2 = As^2\,e^{sx}$, the differential equation [5] becomes

$$Aas^2\,e^{sx} + Abs\,e^{sx} + Ac\,e^{sx} = 0$$

and so equation [6] is a solution if, since e^{sx} is assumed not to be zero,

$$as^2 + bs + c = 0 \qquad [7]$$

This equation is called the *auxiliary equation* or *characteristic equation*. This is a quadratic equation and has roots given by

$$s = \frac{-b \pm \sqrt{b^2 - 4ac}}{2a}$$

Thus if s_1 and s_2 are the two roots, then the general solution is

$$y = A\,e^{s_1 x} + B\,e^{s_2 x} \qquad [8]$$

where A and B are the two constants required for the second-order differential equation.

Example

Determine the general solution of the differential equation

$$\frac{d^2y}{dx^2} - 9y = 0$$

Consider a solution of the form $y = A\,e^{sx}$, then using this value we must have

$$s^2 e^{sx} - 9\,e^{sx} = 0$$

and so $s^2 = 9$. Thus $s = \pm 3$. Hence

$$y = A\,e^{3x} + B\,e^{-3x}$$

Example

Determine the general solution of the differential equation

$$\frac{d^2y}{dx^2} + \frac{dy}{dx} - 2y = 0$$

Consider a solution of the form $y = A\,e^{sx}$, then using this value we must have

$$s^2 e^{sx} + s\,e^{sx} - 2\,e^{sx} = 0$$

Thus we have the auxiliary equation

$$s^2 + s - 2 = 0$$

$$(s - 2)(s + 1) = 0$$

The equation has thus the roots of +2 and −1. Hence the general solution is

$$y = A\,e^{2x} + B\,e^{-x}$$

Review problems

3 Determine the general solutions of the following differential equations:

$$\text{(a)}\ \frac{d^2y}{dx^2} - y = 0,\ \text{(b)}\ \frac{d^2y}{dx^2} - 3\frac{dy}{dx} + 2y = 0,$$

$$\text{(c)}\ \frac{d^2y}{dx^2} + 4\frac{dy}{dx} + 5y = 0$$

6.3.1 The roots of the auxiliary equation

In the above discussion, the roots of the auxiliary equation were two distinct real numbers. There are, however, three possible forms of roots that can occur:

1 Two distinct real roots, i.e. s_1 and s_2 with $s_1 \neq s_2$
2 Two equal real roots, i.e. $s_1 = s_2$
3 Two complex roots, i.e. $s = a \pm jb$

To illustrate the situation with *two equal real roots*, consider the differential equation

$$\frac{d^2y}{dx^2} - 6\frac{dy}{dx} + 9 = 0$$

Using the solution of the form $y = A\,e^{sx}$ gives the auxiliary equation

$$s^2 - 6s + 9 = 0$$

$$(s-3)(s-3) = 0$$

Thus $s_1 = 3$ and $s_2 = 3$, that is the roots are real and equal. Corresponding to the root s_1 we have the solution e^{3x} and corresponding to the root s_2 we have the same solution e^{3x}. This would seem to infer that the general solution is $A\,e^{3x} + B\,e^{3x}$. But this could simply be written as $(A + B)e^{3x}$ or simply $C\,e^{3x}$ where C is $A + B$. Such a solution thus involves only one arbitrary constant and thus cannot be the general solution of the second-order equation, two arbitrary constants being required. To solve the differential equation with two equal roots we need therefore to adopt a different approach. We will take it that one of the solutions is e^{3x} and use a method called the *variation of parameters* to obtain the second solution. This involves transforming the differential equation into another form by using a substitution. Let $y = e^{3x}v$. Then the differentiation of this product gives

$$\frac{dy}{dx} = e^{3x}\frac{dv}{dx} + 3e^{3x}v$$

$$\frac{d^2y}{dx^2} = e^{3x}\frac{d^2v}{dx^2} + 6\,e^{3x}\frac{dv}{dx} + 9\,e^{3x}v$$

Thus substituting these values in the original differential equation gives

$$e^{3x}\frac{d^2v}{dx^2}+6e^{3x}\frac{dv}{dx}+9e^{3x}v-6\left(e^{3x}\frac{dv}{dx}+3e^{3x}v\right)+9e^{3x}v=0$$

Hence

$$e^{3x}\frac{d^2v}{dx^2}=0$$

and so we must have

$$\frac{d^2v}{dx^2}=0$$

A solution of such an equation is $v = x$. Thus, substituting back into $y = e^{3x}v$ gives a solution of the original differential equation as

$$y = x\,e^{3x}$$

Hence the general solution of the differential equation with equal real roots is

$$y = A\,e^{3x} + Bx\,e^{3x} = (A + Bx)e^{3x}$$

To illustrate the situation with *complex roots*, consider the differential equation

$$\frac{d^2y}{dx^2}+2\frac{dy}{dx}+10y=0$$

Using a solution of the form $y = A\,e^{sx}$ gives the auxiliary equation

$$s^2 + 2s + 10 = 0$$

The roots of this equation are thus

$$s=\frac{-2\pm\sqrt{4-40}}{2}$$

$$=-1\pm\sqrt{-9}$$

The square root is of a negative number. If we represent the square root of −1 by j (the symbol generally used by engineers, though mathematicians generally use i), then

$$s=-1\pm\sqrt{(-1)(9)}$$

$$=-1\pm j3$$

The two roots are thus $s_1 = -1 + j3$ and $s_2 = -1 - j3$. Thus the general solution is

$$y = A\,e^{(-1+j3)x} + B\,e^{(-1-j3)x}$$

This equation can be written in another form.

$$y = e^{-x}[A\,e^{+j3x} + B\,e^{-j3x}]$$

$$= e^{-x}[A(\cos 3x + j\sin 3x) + B(\cos 3x - j\sin 3x)]$$

The above equation was arrived at by the use of Euler's equations, namely $e^{jx} = \cos x + j\ \sin x$ and $e^{-jx} = \cos x - j\ \sin x$. Hence

$$y = e^{-x}[(A+B)\cos 3x + j\,(A-B)\sin 3x]$$

Writing $A + B = C$ and $j\,(A - B) = D$ then

$$y = e^{-x}(C\cos 3x + D\sin 3x)$$

In general the solutions of a second-order homogeneous linear differential equation will be of the form:

1 auxiliary equation with real and distinct roots

$$y = A\,e^{s_1 x} + B\,e^{s_2 x} \quad [9]$$

2 auxiliary equation with real and equal roots

$$y = (Ax + B)e^{st} \quad [10]$$

3 auxiliary equation with complex roots $a \pm jb$

$$y = e^{ax}(C\cos bx + D\sin bx) \quad [11]$$

Example

Determine the general solutions of the differential equation

$$\frac{d^2y}{dx^2} + 4\frac{dy}{dx} + 4 = 0$$

With a solution of the form $y = A\,e^{sx}$ then the auxiliary equation will be

$$s^2 + 4s + 4 = 0$$

$$(s+2)(s+2) = 0$$

Thus the two roots are real and equal, both being −2. The general solution will thus be of the form given by equation [10], and so

$$y = (Ax + B)\mathrm{e}^{-2x}$$

Example

Determine the solution of the differential equation

$$\frac{\mathrm{d}^2 y}{\mathrm{d}x^2} - 2\frac{\mathrm{d}y}{\mathrm{d}x} + 5 = 0$$

With a solution of the form $y = A\,\mathrm{e}^{sx}$ the auxiliary equation is

$$s^2 - 2s + 5 = 0$$

This has roots given by

$$s = \frac{2 \pm \sqrt{4 - 20}}{2} = 1 \pm \mathrm{j}2$$

The solution will thus be of the form given by equation [11] and so

$$y = \mathrm{e}^{x}(A\cos 2x + B\sin 2x)$$

Example

Determine the solution of the following differential equation with the given initial conditions.

$$\frac{\mathrm{d}^2 y}{\mathrm{d}x^2} - 6\frac{\mathrm{d}y}{\mathrm{d}x} + 25y = 0, \text{ with } y = 1 \text{ and } \frac{\mathrm{d}y}{\mathrm{d}x} = 7 \text{ at } x = 0$$

With a solution of the form $y = A\,\mathrm{e}^{sx}$ then the auxiliary equation will be

$$s^2 - 6s + 25 = 0$$

Thus

$$s = \frac{6 \pm \sqrt{36 - 100}}{2} = 3 \pm \mathrm{j}4$$

The solution will be of the form given by equation [11] and thus

$$y = \mathrm{e}^{3x}(A\cos 4x + B\sin 4x)$$

Since $y = 1$ when $x = 0$ then

$$1 = 1(A + 0)$$

and so $A = 1$. Differentiating the general solution gives

$$\frac{dy}{dx} = e^{3x}(4A \sin 4x - 4B \cos 4x) + 3e^{3x}(A \cos 4x + B \sin 4x)$$

$$= e^{3x}[(4A + 3B)\sin 4x + (3B - 4A)\cos 4x]$$

Since $dy/dx = 7$ when $x = 0$ then

$$7 = 0[0 + (3A - 4B)]$$

Hence $B = -1$. Thus the solution with the initial conditions is

$$y = e^{3x}(\cos 4x - \sin 4x)$$

Review problems

4 Determine the general solutions of the following differential equations:

(a) $\frac{d^2y}{dx^2} + 2\frac{dy}{dx} + y = 0$, (b) $\frac{d^2y}{dx^2} + 6\frac{dy}{dx} + 9y = 0$,

(c) $\frac{d^2y}{dx^2} - 2\frac{dy}{dx} + 10y = 0$, (d) $\frac{d^2y}{dx^2} - 4\frac{dy}{dx} + 5y = 0$,

(e) $\frac{d^2y}{dx^2} - 2\frac{dy}{dx} + 2y = 0$, (f) $\frac{d^2y}{dx^2} - 9\frac{dy}{dx} + 6y = 0$,

(g) $\frac{d^2y}{dx^2} + 2\frac{dy}{dx} + 4y = 0$, (h) $\frac{d^2y}{dx^2} + 8\frac{dy}{dx} + 16y = 0$

5 Determine the solutions of the following differential equations with the given initial conditions.

(a) $\frac{d^2y}{dx^2} - 2\frac{dy}{dx} + y = 0$, with $y = 0$ and $\frac{dy}{dx} = 3$ at $x = 0$,

(b) $\frac{d^2y}{dx^2} + 6\frac{dy}{dx} + 9y = 0$, with $y = 2$ and $\frac{dy}{dx} = -3$ at $x = 0$,

(c) $\frac{d^2y}{dx^2} + 4\frac{dy}{dx} + 5y = 0$, with $y = 1$ and $\frac{dy}{dx} = -3$ at $x = 0$

6.4 The non-homogeneous linear second-order equation

A non-homogeneous second-order differential equation with constant coefficients a, b and c is of the basic form

$$a\frac{d^2y}{dx^2}+b\frac{dy}{dx}+cy=f(x) \qquad [12]$$

with $f(x)$ being some function of x which is applied to the system. It is often referred to as the forcing function. Such an equation could describe the current oscillations occurring in a circuit containing resistance, capacitance and inductance when there is a voltage source connected to it, $f(x)$ being the function describing this voltage input to the circuit.

As with first-order equations (see section 2.4), linear second-order equations can be solved by considering the solution of the non-homogeneous equation to be equal to the sum of the solution of the corresponding homogeneous equation, the so-called *complementary function* y_c, and another term called the *particular integral* y_p. Thus we can write

$$y=y_c+y_p \qquad [13]$$

The complementary function is the solution of the corresponding homogeneous equation and hence obtained by determining the roots of the auxiliary equations. The particular integral is obtained by considering the form of the forcing function $f(x)$. If this function is a constant then a constant A is tried as the particular integral, if an exponential function then $A\,e^{kx}$ is tried, if of the form $a + bx^2 + cx^3 + \ldots$ then the particular integral tried is $A + Bx^2 + Cx^3 + \ldots$, if it is a sine or cosine then $A\sin\omega x+B\cos\omega x$ is tried. If the forcing function is a sum of some of these terms then the sum of the corresponding terms indicated above is tried, if it is a product then the product is tried. The above suggestions for particular integrals will not work when they are the same form as one of the complementary solutions of the equation. Then it is worthwhile trying the above suggested function multiplied by x or if that is of the form of one of the complementary solutions x^2 or an even higher power of x.

This method of determining the particular integral is called the *method of undetermined coefficients*, since the trial of a form of particular integral with arbitrary coefficients A, B, etc. results in the coefficients being determined. The following examples illustrate this.

Example

Determine the general solution of the differential equation

$$\frac{d^2y}{dx^2}-3\frac{dy}{dx}-4y=4x-5$$

The corresponding homogeneous equation is

$$\frac{d^2y}{dx^2} - 3\frac{dy}{dx} - 4y = 0$$

Trying a solution of the form $y = A\,e^{sx}$ gives the auxiliary equation

$$s^2 - 3s - 4 = 0$$

$$(s+1)(s-4) = 0$$

Thus

$$y_c = A\,e^{-x} + B\,e^{4x}$$

For the particular integral we can try a function of the form $ax + b$. This gives $dy/dx = a$ and $d^2y/dx^2 = 0$. Thus, using these values gives

$$0 - 3a - 4(ax + b) = 4x - 5$$

Equating coefficients of x gives $a = -1$ and equating integers gives $b = 2$. Hence

$$y_p = -x + 2$$

Thus the general solution is

$$y = y_c + y_p = A\,e^{-x} + B\,e^{4x} - x + 2$$

Example

Determine the solution of the differential equation

$$\frac{d^2y}{dx^2} - 2\frac{dy}{dx} - 3y = 4\,e^{3x}$$

For the corresponding homogeneous equation, namely,

$$\frac{d^2y}{dx^2} - 2\frac{dy}{dx} - 3y = 0$$

we can try a complementary function of the form $A\,e^{st}$. This gives the auxiliary equation

$$s^2 - 2s - 3 = 0$$

$$(s+1)(s-3) = 0$$

Thus the complementary function is

$$y_c = A\,e^{-x} + B\,e^{3x}$$

For the particular integral, since the forcing function is $4\,e^{3x}$, it would seem worth trying $A\,e^{3x}$ (note that this is of the same form as a solution of the homogeneous equation). But when we use this value in the differential equation then

$$9A\,e^{3x} - 6A\,e^{3x} - 3A\,e^{3x} = 4\,e^{3x}$$

$$0 = 4\,e^{3x}$$

There is no real value of x for which this can be true. Thus we need to try another form of particular integral. Try $Ax\,e^{3x}$. Using this value in the differential equation gives

$$(9Ax\,e^{3x} + 6A\,e^{3x}) - 2(3Ax\,e^{3x} + A\,e^{3x}) - 3Ax\,e^{3x} = 4\,e^{3x}$$

$$0x\,e^{3x} + 4A\,e^{3x} = 4\,e^{3x}$$

Hence $A = 1$ and so the particular integral is

$$y_p = x\,e^{3x}$$

The general solution for the differential equation is thus

$$y = y_c + y_p = A\,e^{-x} + B\,e^{3x} + x\,e^{3x}$$

Note that when the particular integral is of the same form as a solution of the homogeneous equation, it will not work and a particular integral of a different form must be used.

Example

Determine the general solution of the differential equation

$$\frac{d^2y}{dx^2} - 4\frac{dy}{dx} + 4y = 4x + 8\cos 2x$$

For the homogeneous form of the above equation, trying a solution of the form $A\,e^{sx}$ gives an auxiliary equation of

$$s^2 - 4s + 4 = 0$$

$$(s-2)(s-2) = 0$$

Thus, since there are equal real roots, the complementary function is

$$y_c = e^{2x}(Ax + B)$$

The particular integral of the $4x$ term is $a + bx$, the particular integral for the cosine term $c \cos 2x + d \sin 2x$. Thus the particular integral which can be tried is $a + bx + c \cos 2x + d \sin 2x$. Using this value the differential equation then becomes

$$(-4c\cos 2x - 4d\sin 2x) - 4(b - 2c\sin 2x + 2d\cos 2x) \\ + 4(a + bx + c\cos 2x + d\sin 2x) = 4x + 8\cos 2x$$

Equating constant terms gives $-4b + 4a = 0$ and thus $a = b$. Equating the coefficients of x gives $4b = 4$ and so $b = 1$, and hence we have $a = 1$. Then equating the coefficients of cos $2x$ gives $-4c - 8d + 4c = 8$ and so $d = -1$. Equating the coefficients of $\sin 2x$ gives $-4d + 8c + 4d = 0$ and so $c = 0$. Thus the particular integral is

$$y_p = 1 + x - \sin 2x$$

and the general solution is

$$y = y_c + y_p = e^{2x}(Ax + B) + 1 + x - \sin 2x$$

Review problems

6 Determine the general solutions of the following differential equations:

$$\text{(a) } \frac{d^2y}{dx^2} + 4\frac{dy}{dx} + 3y = 9, \text{ (b) } \frac{d^2y}{dx^2} + 4y = 8,$$

$$\text{(c) } \frac{d^2y}{dx^2} + 4\frac{dy}{dx} = 8, \text{ (d) } \frac{d^2y}{dx^2} - 4y = 8e^{2x},$$

$$\text{(e) } \frac{d^2y}{dx^2} + 4y = 24x^3, \text{ (f) } \frac{d^2y}{dx^2} - 4\frac{dy}{dx} + 3y = 12x - 4,$$

$$\text{(g) } \frac{d^2y}{dx^2} - \frac{dy}{dx} - 2y = 10\cos x,$$

$$\text{(h) } \frac{d^2y}{dx^2} - 2\frac{dy}{dx} - 3y = 4e^{x} - 10\sin x,$$

$$\text{(i) } \frac{d^2y}{dx^2} - 3\frac{dy}{dx} + 2y = 6e^{-x} - 13\cos 3x$$

6.5 Higher order equations

Consider an n-th order homogeneous linear differential equation with constant coefficients. Earlier in this chapter we found that such an equation when second order with an auxiliary equation with two distinct real roots s_1 and s_2 has a general solution

$$y = A\,e^{s_1 x} + B\,e^{s_2 x}$$

An n-th order equation with n distinct real roots s_1, s_2, ...s_n has a general solution

$$y = A\,e^{s_1 x} + B\,e^{s_2 x} + C\,e^{s_3 x} + \dots \text{ for } n \text{ terms} \qquad [14]$$

with A, B, C, ... being arbitrary constants.

Thus, for example, for the third-order differential equation

$$\frac{d^3y}{dx^3} - 4\frac{d^2y}{dx^2} + \frac{dy}{dx} + 6y = 0$$

the auxiliary equation obtained by trying $y = A\,e^{st}$ is

$$s^3 - 4s^2 + s + 6 = 0$$

$$(s+1)(s-2)(s-3) = 0$$

Thus there are three distinct real roots and so the general solution is

$$y = A\,e^{-x} + B\,e^{2x} + C\,e^{3x}$$

If the n-th order differential equation has an auxiliary equation with two identical real roots s then the general solution would include terms e^{sx} and $x\,e^{sx}$, if three identical real roots e^{sx}, $x\,e^{sx}$ and $x^2\,e^{sx}$. Consider, for example, the differential equation

$$\frac{d^4y}{dx^4} - 5\frac{d^3y}{dx^3} + 6\frac{d^2y}{dx^2} + 4\frac{dy}{dx} - 8y = 0$$

This has the auxiliary equation

$$s^4 - 5s^3 + 6s^2 + 4s - 8 = 0$$

$$(s-2)(s-2)(s-2)(s+1) = 0$$

Thus the general solution is

$$y = A\,e^{2x} + Bx\,e^{2x} + Cx^2\,e^{2x} + D\,e^{-x}$$

If the n-th order homogeneous differential equation has an auxiliary equation with pairs of complex roots ($a \pm jb$), neither of which is repeated, then the corresponding part of the general solution will be

$$y = e^{ax}(A \sin bx + B \cos bx) \qquad [15]$$

If the complex roots are repeated then the constants multiplying the sine and cosine terms become of the form $A + Bx + Cx^2 + \dots$ Thus, for example, for a homogeneous differential equation giving an auxiliary equation with the roots

$$(1 + j3)(1 - j3)(1 + j3)(1 - j3)$$

then the general solution is

$$y = e^{x}[(A + Bx)\sin 3x + (C + Dx)\cos 3x]$$

Consider now a non-homogeneous n-th order differential equation with constant coefficients. The method described earlier in this chapter of considering the general solution to be the sum of the complementary function and the particular integral can be used. The complementary function is the solution of the corresponding homogeneous equation and found as described above; the particular integral can be found as described earlier by the method of undetermined coefficients. This method of undetermined coefficients can be used whenever the forcing function is a function, or a linear combination of products of functions, which is or are polynomials in x, exponential functions of x, or sines or cosines of x. For example, the forcing function $2x^2 + 3 \sin 5x$ would enable the method to be used, but the function $\tan x$ or $1/x$ would not. For such functions another method, the variation of parameters (not discussed in this book), can be used to determine the particular integral.

Example

Determine the general solution of the following fourth-order differential equation:

$$\frac{d^4y}{dx^4} + 3\frac{d^3y}{dx^3} + 3\frac{d^2y}{dx^2} + \frac{dy}{dx} = 0$$

The auxiliary equation for the differential equation is

$$s^4 + 3s^3 + 3s^2 + s = 0$$

$$s(s + 1)(s + 1)(s + 1) = 0$$

Thus the roots have the values 0 and the three times repeated root of -1. The general solution is thus

$$y = A\,e^0 + B\,e^{-x} + Cx\,e^{-x} + Dx^2\,e^{-x}$$

Example

Determine the general solution of the differential equation

$$\frac{d^4y}{dx^4} - 2\frac{d^3y}{dx^3} + 5\frac{d^2y}{dx^2} - 8\frac{dy}{dx} + 4y = 20\,e^x$$

The auxiliary equation for the corresponding homogeneous equation is

$$s^4 - 2s^3 + 5s^2 - 8s + 4 = 0$$

$$(s-1)(s-1)(s^2+4) = 0$$

The roots are thus 1, 1 and the pair of complex roots given for $s^2 + 4$, i.e. $s = \pm j2$. Thus the complementary function is

$$y_c = e^x(Ax + B) + e^0(C\cos 2x + D\sin 2x)$$

The particular integral will be $x^2\,e^x$, since there are two terms involving e^x in the complementary function. Hence, using this particular integral in the differential equation gives

$$E(x^2\,e^x + 8x\,e^x + 12\,e^x) - 2E(x^2\,e^x + 6x\,e^x + 6\,e^x)$$
$$+ 5E(x^2\,e^x + 4x\,e^x + 2\,e^x) - 8E(x^2\,e^x + 2x\,e^x) + 4Ex^2\,e^x = 20\,e^x$$

Hence, $10E\,e^x = 20\,e^x$ and so $E = 2$. Thus the particular integral is $2x^2\,e^x$. The general solution is thus

$$y = e^x(Ax + B) + e^0(C\cos 2x + D\sin 2x) + 2x^2\,e^x$$

Review problems

7 Determine the general solutions of the following differential equations:

(a) $\dfrac{d^3y}{dx^3} + 2\dfrac{d^2y}{dx^2} + \dfrac{dy}{dx} = 0$, (b) $\dfrac{d^4y}{dx^4} + 4\dfrac{d^2y}{dx^2} = 0$,

(c) $\dfrac{d^3y}{dx^3} - \dfrac{d^2y}{dx^2} - 6\dfrac{dy}{dx} = x^2 - 3x - 2$,

(d) $\dfrac{d^3y}{dx^3} + 3\dfrac{d^2y}{dx^2} + 3\dfrac{dy}{dx} + y = x + 6$

Further problems

8 Solve the following differential equations by two stages of integration.

(a) $\dfrac{d^2y}{dx^2} = 2$, with $y = 0$ and $\dfrac{dy}{dx} = 1$ at $x = 0$,

(b) $\dfrac{d^2y}{dx^2} = 4x$, with $y = 1$ and $\dfrac{dy}{dx} = 2$ at $x = 0$,

(c) $\dfrac{d^2y}{dx^2} = 2x + 1$, with $y = 0$ and $\dfrac{dy}{dx} = 0$ at $x = 0$

9 The deflection y of a simply supported beam of length L at a distance x from one end with a uniformly distributed load of w per unit length is given by

$$EI\frac{d^2y}{dx^2} = -\frac{wLx}{2} + \frac{wx^2}{2}$$

At the mid span, when $x = \frac{1}{2}L$, then $dy/dx = 0$. At $x = 0$ we also have $y = 0$. Solve the differential equation.

10 The deflection of a cantilever of length L at a distance x from the fixed end when supporting a uniformly distributed load of w per unit length is given by

$$EI\frac{d^2y}{dx^2} = \tfrac{1}{2}w(L - x)^2$$

with $y = 0$ and $dy/dx = 0$ at $x = 0$. Solve the differential equation.

11 Determine the general solutions of the following differential equations:

(a) $\dfrac{d^2y}{dx^2} - 4y = 0$, (b) $\dfrac{d^2y}{dx^2} - \dfrac{dy}{dx} = 0$,

(c) $\dfrac{d^2y}{dx^2} - 6\dfrac{dy}{dx} + 5y = 0$, (d) $\dfrac{d^2y}{dx^2} + 4\dfrac{dy}{dx} + 4y = 0$,

(e) $\dfrac{d^2y}{dx^2} + 3\dfrac{dy}{dx} - 10y = 0$, (f) $\dfrac{d^2y}{dx^2} + 6\dfrac{dy}{dx} + 9y = 0$,

(g) $\dfrac{d^2y}{dx^2} - 6\dfrac{dy}{dx} + 25y = 0$, (h) $\dfrac{d^2y}{dx^2} + 4\dfrac{dy}{dx} + 5y = 0$

12 Determine the solutions of the following differential equations with the given initial conditions.

(a) $\dfrac{d^2y}{dx^2} + 9y = 0$, with $y = 0$ and $\dfrac{dy}{dx} = 1$ at $x = 0$,

(b) $\dfrac{d^2y}{dx^2} - 4\dfrac{dy}{dx} + 4y = 0$, with $y = 3$ and $\dfrac{dy}{dx} = 1$ at $x = 0$,

(c) $\dfrac{d^2y}{dx^2} - 2\dfrac{dy}{dx} + 10y = 0$, with $y = 2$ and $\dfrac{dy}{dx} = 1$ at $x = 0$

13 Determine the general solutions of the following differential equations:

(a) $\dfrac{d^2y}{dx^2} - 3\dfrac{dy}{dx} + 2y = 2\,e^{-2x}$,

(b) $\dfrac{d^2y}{dx^2} - 3\dfrac{dy}{dx} + 2y = 2\,e^{2x}$,

(c) $\dfrac{d^2y}{dx^2} - 3\dfrac{dy}{dx} + 2y = 2 + 4x^2$,

(d) $\dfrac{d^2y}{dx^2} - 3\dfrac{dy}{dx} + 2y = 20\sin 2x$,

(e) $\dfrac{d^2y}{dx^2} + \dfrac{dy}{dx} - 6y = 4\sin 2x$,

(f) $\dfrac{d^2y}{dx^2} - 3\dfrac{dy}{dx} - 10y = 4 - e^{-2x}$,

(g) $\dfrac{d^2y}{dx^2} - 5\dfrac{dy}{dx} + 6y = -3\sin x$,

(h) $\dfrac{d^2y}{dx^2} - 2\dfrac{dy}{dx} + 2x = e^{x}\sin x$

14 The following differential equation describes the motion of a body. Solve the equation for $y = 0$ and $dy/dt = 0$ at $t = 0$.

$$\frac{d^2y}{dx^2} + \omega_0^2 x = k\cos\omega t$$

15 An object of mass 5 kg is moving in a straight line along the x axis. It is subject to a resisting force of $(20v + 15x)$ N, with v being in m/s and x in m. Derive the differential equation relating the displacement of the object with time and then solve it for the initial conditions $x = 3$, $v = 1$ at $t = 0$.

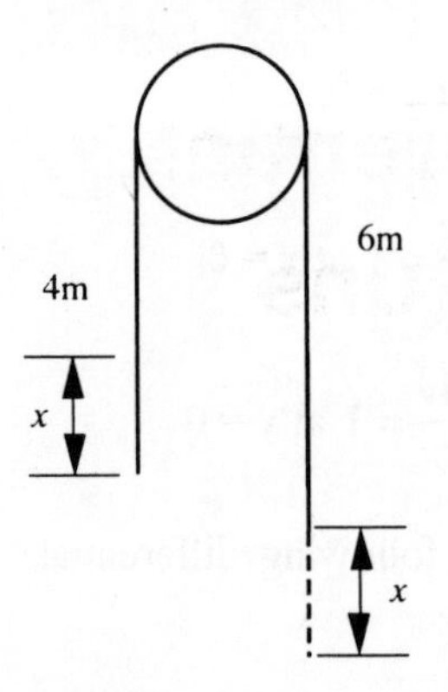

Fig. 6.1 Problem 16

16 A 10 m length of chain hangs over a smooth rod such that at time $t = 0$ there is 4 m on one side and 6 m on the other, as illustrated in figure 6.1. Taking the acceleration due to gravity to be 10 m/s^2, derive the differential equation relating the displacement x from this initial position and the time t. Hence solve the equation for $x = 0$ and $v = 0$ at $t = 0$. Hint: take the mass of the total length of chain to be m and then consider for x the excess amount of chain on one side which will be causing a gravitational force to act on the chain, making it accelerate.

17 An electric circuit gives an output voltage v which is given by the differential equation

$$\frac{d^2v}{dt^2} + 6\frac{dv}{dt} + 25v = V - Ve^{-t}$$

Derive the solution given that $v = 0$ and $dv/dt = 0$ at $t = 0$.

18 Determine the general solutions of the following differential equations:

(a) $\dfrac{d^3y}{dx^3} - 2\dfrac{d^2y}{dx^2} - 7\dfrac{dy}{dx} - 4\dfrac{dy}{dx} = 0$,

(b) $\dfrac{d^3y}{dx^3} - 9\dfrac{d^2y}{dx^3} + 27\dfrac{dy}{dx} - 27y = 0$,

(c) $\dfrac{d^3y}{dx^3} - 6\dfrac{d^2y}{dx^2} + 12\dfrac{dy}{dx} - 8\dfrac{dy}{dx} = 0$,

(d) $\dfrac{d^3y}{dx^3} - 3\dfrac{d^2y}{dx^2} + 3\dfrac{dy}{dx} - y = e^{-x}$,

(e) $\dfrac{d^3y}{dx^3} - 3\dfrac{d^2y}{dx^2} + 4y = 4e^{x} - 18e^{-x}$

7 Second-order differential equations: oscillations

7.1 Undamped mechanical oscillations

As an example of a mechanical system, consider a vertically suspended spring with a mass m attached to its lower end, as illustrated by figure 7.1. The mass is pulled down some distance x and then released. When a spring, which obeys Hooke's law, is stretched then the force F exerted by the spring is proportional to the extension, i.e. $F = kx$ with k being a constant called the *spring stiffness*. Thus the restoring force exerted on the mass as a result of the spring being stretched is an upward-directed force of kx. Because the force is acting in an upward direction and x increases as the spring is stretched in a downward direction, the restoring force is usually written as $-kx$. This force causes an acceleration a of the mass. Thus, using Newton's second law, $-kx = ma$. The acceleration is proportional to the displacement, the direction of the acceleration being in the opposite direction to that in which x increases. Such an oscillation is said to be *simple harmonic motion*. Since acceleration is the rate of change of velocity with time and velocity is the rate of change of the displacement x with time, then

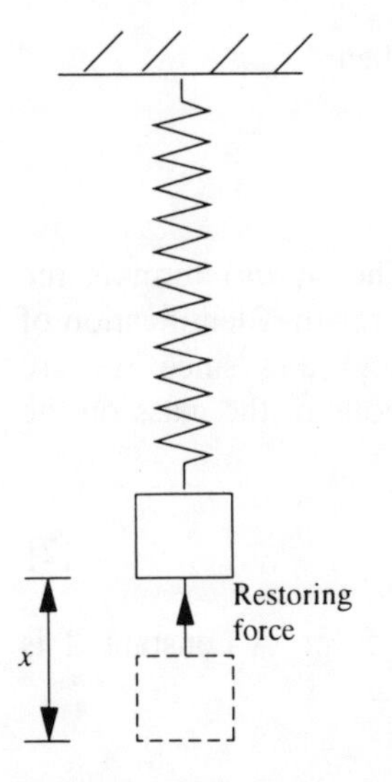

Fig. 7.1 Restoring force on mass

$$m\frac{d^2x}{dt^2} = -kx$$

or

$$m\frac{d^2x}{dt^2} + kx = 0 \qquad [1]$$

We can solve this homogeneous differential equation by considering a solution of the form $x = A\,e^{st}$. Thus the auxiliary equation is

$$ms^2 + k = 0$$

Hence

$$s = \pm\sqrt{-\frac{k}{m}} = \pm \mathrm{j}\sqrt{\frac{k}{m}}$$

where j is the square root of minus 1. Thus the general solution is

$$x = A\,\mathrm{e}^{\mathrm{j}\sqrt{k/m}\,t} + B\,\mathrm{e}^{-\mathrm{j}\sqrt{k/m}\,t}$$

This equation can be written in another form by using Euler's equations, namely $\mathrm{e}^{\mathrm{j}y} = \cos y + \mathrm{j}\sin y$ and $\mathrm{e}^{-\mathrm{j}y} = \cos y - \mathrm{j}\sin y$.

$$x = A\cos\left(\sqrt{k/m}\right)t + \mathrm{j}A\sin\left(\sqrt{k/m}\right)t$$
$$+B\cos\left(\sqrt{k/m}\right)t - \mathrm{j}B\sin\left(\sqrt{k/m}\right)t$$

$$x = (A+B)\cos\left(\sqrt{k/m}\right)t + \mathrm{j}(A-B)\sin\left(\sqrt{k/m}\right)t$$

Writing for $(A+B) = C$ and $\mathrm{j}(A-B) = D$, then

$$x = C\cos\left(\sqrt{k/m}\right)t + D\sin\left(\sqrt{k/m}\right)t$$

This equation describes an oscillation. The $\sqrt{(k/m)}$ term is the angular frequency (see later in this section for the identification of angular frequency as $2\pi \times$ the frequency) and since we are concerned with the natural or free oscillations of the mass on the spring it is designated as ω_n. Thus

$$x = C\cos\omega_n t + D\sin\omega_n t \qquad [2]$$

This equation is often written in another form. A constant A is chosen such that

$$A^2 = C^2 + D^2$$

We can think of A being the hypotenuse of a right-angled triangle and C and D the other sides, as in figure 7.2. Then, for that triangle $\cos\alpha = C/A$ and $\sin\alpha = D/A$. Equation [2] can thus be written as

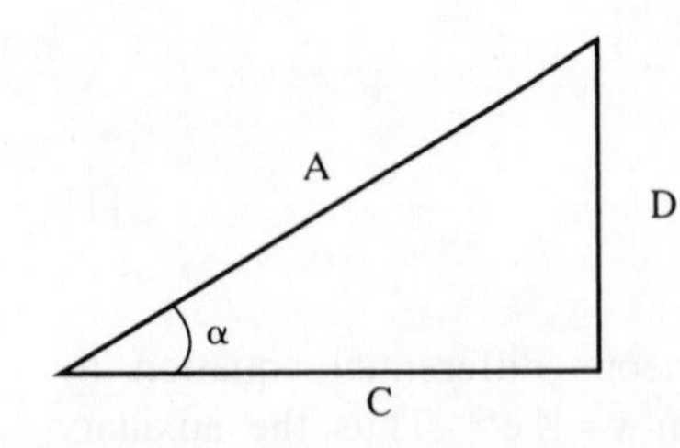

Fig. 7.2 Phase angle α

$$x = A\cos\omega_n t\cos\alpha + A\sin\omega_n t\sin\alpha$$

But $\cos(\omega_n t - \alpha) = \cos\omega_n t\cos\alpha + \sin\omega_n t\sin\alpha$. Thus

$$x = A\cos(\omega_n t - \alpha) \qquad [3]$$

The constants A and α both have physical significance. A is the maximum value that x can have and is thus the *amplitude* of the oscillation. α is called the *phase angle*. Figure 7.3 illustrates these terms on a graph of equation [3]. We can think of the graph of $\cos \omega_n t$ being generated by a line rotating with an angular velocity ω_n, the initial position of the line at time $t = 0$ being vertical. The graph of $\cos(\omega_n t - \alpha)$ is then generated by the line starting its rotation at an angle α which is behind the vertical starting position for $\cos \omega_n t$. If we had $\cos(\omega_n t + \alpha)$ then the line would start an angle α ahead of the vertical starting position for $\cos \omega_n t$.

Note that equations [2] and [3] are the general solutions. In engineering textbooks a solution is often quoted for the condition that $x = 0$ at $t = 0$, namely

$$x = A \cos \omega_n t \qquad [4]$$

This is equation [3] with $\alpha = 0$.

The *periodic time* T of an oscillation is the time taken to complete one oscillation. Thus, for the line in figure 7.3 rotating with angular velocity ω_n, since there are 2π radians in one complete revolution then

$$T = \frac{2\pi}{\omega_n} \qquad [5]$$

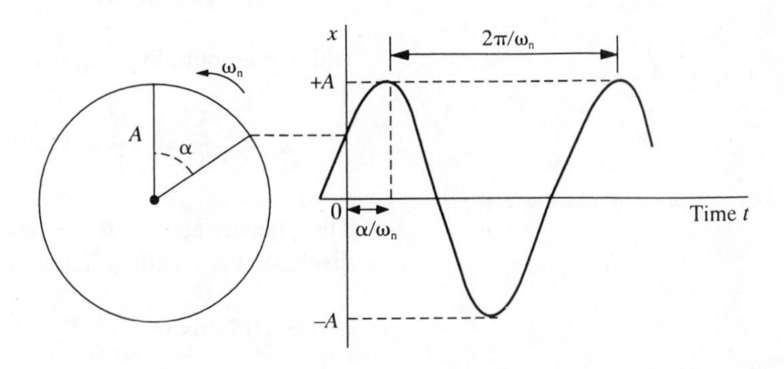

Fig. 7.3 $x = A \cos(\omega_n t - \alpha)$

The *frequency* f of an oscillation is the number of complete cycles per second and is thus $1/T$. Thus

$$f = \frac{\omega_n}{2\pi} \qquad [6]$$

Thus equation [3] written in terms of frequency is

$$x = A \cos(2\pi f_n t - \alpha)$$

Note that the differential equation describing the oscillation, i.e. equation [1], is often written in terms of the natural angular frequency ω_n as

$$\frac{d^2x}{dt^2} + \omega_n^2 x = 0 \qquad [7]$$

Example

An object of mass 1 kg is attached to the end of a vertically suspended spring, the spring having a stiffness of 9 N/m. The mass is pulled down a distance of 0.2 m and then released. Determine the differential equation for the motion and the position of the object after a time of 0.1 s.

Equation [1] gives for the differential equation

$$m\frac{d^2x}{dt^2} + kx = 0$$

$$\frac{d^2x}{dt^2} + 9x = 0$$

The solution of this differential equation is given by equation [3] as

$$x = A\cos(\omega_n t - \alpha)$$

with the amplitude being 0.2 m and

$$\omega_n = \sqrt{\frac{k}{m}} = \sqrt{\frac{9}{1}} = 3 \text{ rad/s}$$

Thus, since at $t = 0$ the oscillation is starting at a maximum displacement (with no initial velocity) then the phase angle is zero,

$$x = 0.2\cos 3t$$

Thus for $t = 0.1$ s then $3t = 0.3$ rad $= 17.2°$ and so $x = 0.191$ m.

Example

An object of mass 1 kg is attached to the end of a vertically suspended spring, the spring having a stiffness of 9 N/m. The mass is pulled down a distance of 0.2 m and then released with an initial velocity in the upwards direction of 0.1 m/s. Determine the differential equation for the motion and the position of the object after a time of 0.1 s.

This example only differs from the previous one in that the object starts at time $t = 0$ with an initial velocity of -0.1 m/s, the minus sign being because the direction of the velocity is in the opposite direction to that in which x increases. The differential equation will be the same and the solution is

$$x = A\cos(\omega_n t - \alpha)$$

with the phase angle not being zero in this instance. Since we know the velocity at $t = 0$ we can determine the phase angle by differentiating the above equation.

$$v = \frac{dx}{dt} = -\omega_n A \sin(\omega_n t - \alpha)$$

Hence

$$-0.1 = -3 \times 0.2 \sin(-\alpha)$$

Thus the phase angle is $-9.6°$. Hence

$$x = A\cos(\omega_n t + 9.6°)$$

At $t = 0.1$ s then $\omega_n t = 0.3$ rad $= 17.2°$ and so

$$x = 0.2\cos(17.2° + 9.6°) = 0.18 \text{ m}$$

Alternatively, the solution can be taken in the form given by equation [2], i.e.

$$x = C\cos\omega_n t + D\sin\omega_n t$$

At $t = 0$ we have $x = 0.2$ and so putting these values in the equation gives $C = 0.2$. Differentiating the equation gives

$$v = \frac{dx}{dt} = -\omega_n C\sin\omega_n t + \omega_n D\cos\omega_n t$$

Thus, since $v = -1$ m/s at $t = 0$ then $-0.1 = 3D$. Thus

$$x = 0.2\cos 3t - 0.033\sin 3t$$

When $t = 0.1$ s then this gives $x = 0.18$ m.

Review problems

1 An object has a motion which is described by the differential equation

$$\frac{d^2x}{dt^2} + 16x = 0$$

What is (a) the natural angular frequency, (b) the displacement after a time of 0.1 s if at $t = 0$ there is a displacement x of 3 m and a velocity of 16 m/s in the direction of x increasing?

2 An object oscillates with a displacement x which varies with time according to the differential equation

$$\frac{d^2x}{dt^2} + 1x = 0$$

Solve the equation for when the object at an initial displacement of 1 m has (a) zero initial velocity, (b) an initial velocity of 1 m/s in the direction of x increasing, (c) an initial velocity of 1 m/s in the opposite direction to that in which x increases.

7.1.1 Rotational oscillations

Consider a system with a moment of inertia I when it is rotated through some angle θ, perhaps a shaft being twisted through this angle. If the rotational stiffness of the system is q, i.e. the torque needed to rotate through 1 rad, then the system will exert a restoring torque of $-q\theta$. When the system is released then the restoring torque will result in an angular acceleration of α. Thus applying the rotational version of Newton's second law

$$\text{torque} = I\alpha = -q\theta$$

But the angular acceleration is the rate of change of angular velocity with time and the angular velocity is the rate of change of angular displacement with time. Thus

$$I\frac{d^2\theta}{dt^2} + q\theta = 0 \qquad [8]$$

This equation is of the same form as equation [1] for the mass oscillating on the spring and will have a similar form of solution, namely

$$\theta = \theta_{max} \cos(\omega_n t - \alpha) \qquad [9]$$

with $\omega_n = \sqrt{(q/I)}$ The periodic time T of the oscillation is $2\pi/\omega_n$ and the frequency is $1/T$.

7.1.2 Simple harmonic motion

Consider an oscillation described by equation [3], or equation [4].

$$x = A\cos(\omega_n t - \alpha)$$

Then the velocity v is $\mathrm{d}x/\mathrm{d}t$ and so is

$$v = \frac{\mathrm{d}x}{\mathrm{d}t} = -\omega_n A \sin(\omega_n t - \alpha)$$

The acceleration a is $\mathrm{d}v/\mathrm{d}t$ and so is

$$a = \frac{\mathrm{d}v}{\mathrm{d}t} = -\omega_n^2 A \cos(\omega_n t - \alpha) = -\omega_n^2 x \qquad [10]$$

This relation between the acceleration and the displacement defines *simple harmonic motion*. Since the periodic time T is $2\pi/\omega$ (equation [5]) then, for simple harmonic motion, equation [10] gives

$$T = 2\pi\sqrt{\frac{x}{a}} \qquad [11]$$

A similar line of reasoning can be used with rotational motion to derive the relationship

$$\alpha = -\omega_n^2 \theta \qquad [12]$$

The periodic time T for rotational simple harmonic motion is thus

$$T = 2\pi\sqrt{\frac{\theta}{\alpha}} \qquad [13]$$

Example

A horizontal platform, supporting an instrument, rests on four vertical springs. If the springs each have a stiffness of k and the mass of the platform plus instrument is m, derive the differential equation and an equation for the periodic time with which the platform plus instrument naturally oscillates.

The total restoring force when the platform is displaced vertically is the sum of the forces acting on each spring and is thus $4kx$. Thus, according to Newton's second law, we must have $ma = -4kx$. This indicates that the motion is simple harmonic since we have $a \propto -x$. The differential equation for the oscillation of the system when it is displaced is thus

$$m\frac{\mathrm{d}^2 x}{\mathrm{d}t^2} + 4kx = 0$$

Thus, if we put this equation into the form of equation [7] we have $\omega_n^2 = 4k/m$ and thus

$$T = 2\pi\sqrt{\frac{m}{4k}}$$

Alternatively, we could have derived this equation by just using the relationship $ma = -4kx$ and substituting for x/a in equation [11].

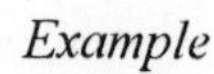

Example

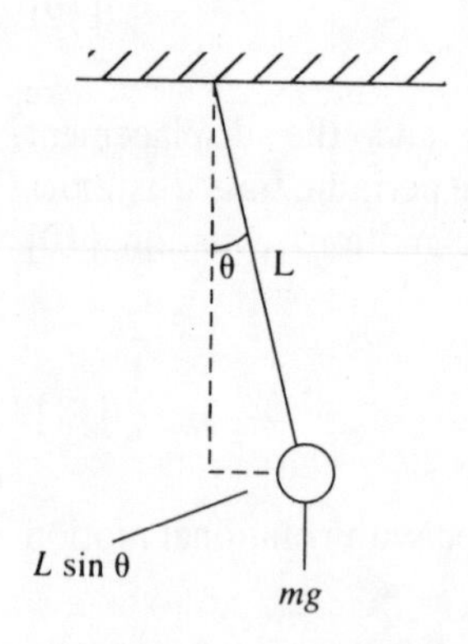

Fig. 7.4 Simple pendulum

A simple pendulum consists of a concentrated mass m which swings on the end of a massless string of length L (figure 7.4). Determine the periodic time with which the pendulum naturally oscillates when given an initial angular displacement.

The restoring torque is $-mgL\sin\theta$. Thus, using the angular version of Newton's second law,

$$-mgL\sin\theta = I\alpha$$

The moment of inertia I about the axis of suspension is mL^2 and if the motion is restricted to small angular displacements then $\sin\theta \approx \theta$ and so

$$\alpha = -\frac{g\theta}{L}$$

The motion is thus simple harmonic since $\alpha \propto -\theta$. Hence, using equation [13],

$$T = 2\pi\sqrt{\frac{\theta}{\alpha}} = 2\pi\sqrt{\frac{L}{g}}$$

Review problems

3 An object of mass m is suspended from a spring of stiffness k_1 which is in turn suspended from a spring of stiffness k_2. Derive an equation for the periodic time of the oscillation.

4 Figure 7.5 shows a horizontal bar of mass m and length L which is pivoted at one end and supported at the other end by a vertical spring of stiffness k from a rigid support. At equilibrium the bar is horizontal. Derive an equation for the periodic time of natural oscillations of the system.

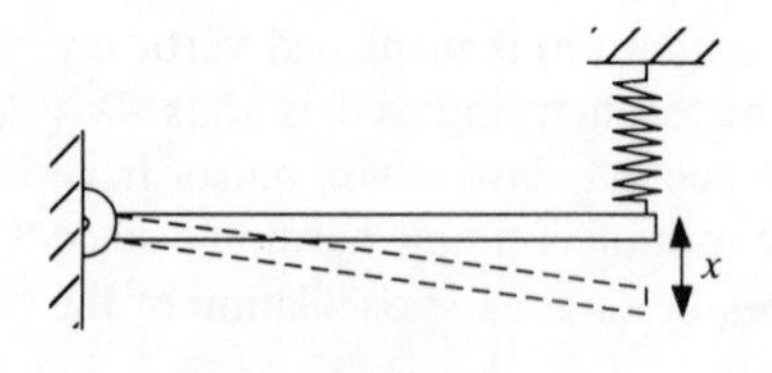

Fig. 7.5 Example

7.2 Damped mechanical oscillations

In the discussions in section 7.1, no account was taken of any frictional or damping forces acting on an oscillating mass. We can consider the basic mechanical oscillation model of a mass on a spring, considered in figure 7.1, as being modified by the inclusion of *viscous damping*. Such damping is represented by the movement of a piston in a container, i.e. a dashpot, in the way shown in figure 7.6. Viscous damping produces a resistive force which is proportional to the velocity.

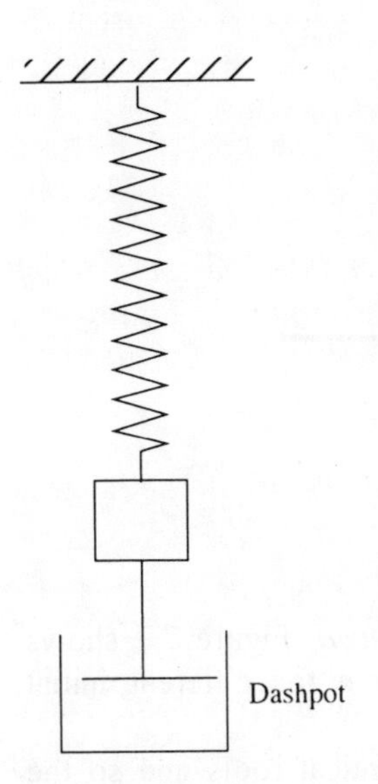

Fig. 7.6 Damped system

When the mass m on the spring is pulled down a distance x then the restoring force exerted by the spring on the mass is $-kx$, where k is the spring stiffness. When the mass is released and moves under the action of this restoring force then because it acquires a velocity there will be a resistive damping force. Thus the resultant force acting on the mass at some instant is the restoring force minus the damping force. The damping force is proportional to the velocity, i.e. $\mathrm{d}x/\mathrm{d}t$. If c is the constant of proportionality, i.e. the damping force per unit velocity, then

$$\text{resultant force} = -kx - c\frac{\mathrm{d}x}{\mathrm{d}t}$$

This resultant force causes an acceleration and thus Newton's second law gives

$$m\frac{\mathrm{d}^2x}{\mathrm{d}t^2} = -kx - c\frac{\mathrm{d}x}{\mathrm{d}t}$$

$$m\frac{\mathrm{d}^2x}{\mathrm{d}t^2} + c\frac{\mathrm{d}x}{\mathrm{d}t} + kx = 0 \qquad [14]$$

Following the method of solution described in section 6.3 for the homogeneous second-order differential equation, then if we try a solution of the form $x = A\,\mathrm{e}^{st}$ the following auxiliary equation is produced:

$$ms^2 + cs + k = 0$$

The roots of this quadratic equation are given by

$$s = \frac{-c \pm \sqrt{c^2 - 4mk}}{2m} \qquad [15]$$

The form of the solution will depend on the value of the square root term.

1 With $c^2 > 4mk$ there are two distinct real roots s_1 and s_2 and so the solution is (as given in equation [9], chapter 6)

$$x = A\,\mathrm{e}^{s_1 t} + B\,\mathrm{e}^{s_2 t} \qquad [16]$$

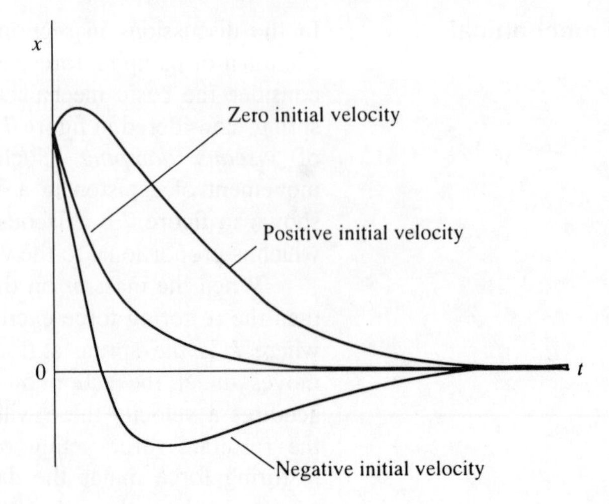

Fig. 7.7 Overdamped

This situation is described as *overdamped*. Figure 7.7 shows how the displacement varies with time for different initial velocities at time $t = 0$.

2 With $c^2 = 4mk$ there are two real identical roots and so the solution is (as given in equation [10], chapter 6)

$$x = (At + B)e^{st} \qquad [17]$$

This situation is described as *critically damped*. Figure 7.8 shows how the displacement varies with time for different initial velocities at time $t = 0$. As $t \to \infty$ then the displacement tends to zero.

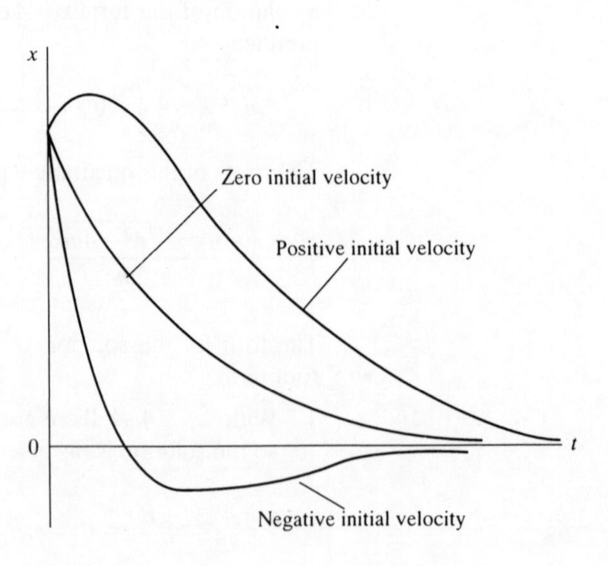

Fig. 7.8 Critically damped

3 With $c^2 < 4mk$ there are a pair of complex roots. These roots are

$$s = -\frac{c}{2m} \pm \mathrm{j}\frac{\sqrt{4mk - c^2}}{2m}$$

This can be written as

$$s = -\mu \pm \mathrm{j}\omega$$

where $\mu = c/2m$ and $\omega = \sqrt{4mk - c^2}/2m$. The solution is (as given in equation [11], chapter 6)

$$x = \mathrm{e}^{-\mu t}(C\cos\omega t + D\sin\omega t) \qquad [18]$$

This can also be written as

$$x = A\,\mathrm{e}^{-\mu t}\cos(\omega t - \alpha) \qquad [19]$$

This situation is described as *underdamped*. The equation describes an oscillation with an amplitude of $A\,\mathrm{e}^{-\mu t}$, i.e. an amplitude which decreases with time. Figure 7.9 shows how the displacement varies with time for an initial velocity of zero

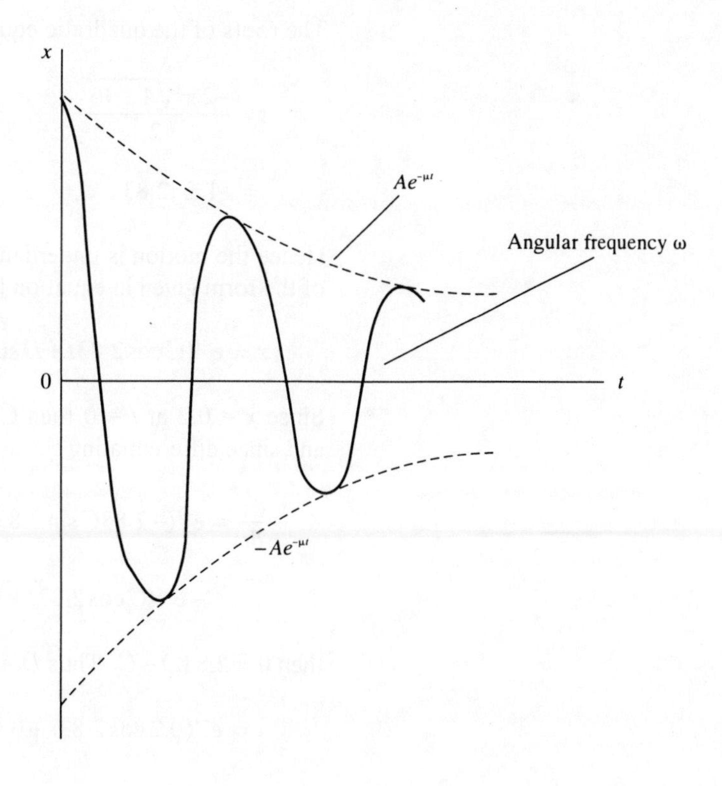

Fig. 7.9 Underdamped

at time $t = 0$, i.e. with no phase angle α. The damped oscillation has an angular frequency of ω. As c decreases to zero so ω tends to $\sqrt{(4mk)}/2m$, i.e. ω_n.

Example

Determine the differential equation and its solution for a system consisting of a mass of 1 kg suspended from a spring of stiffness 9 N/m and subject to damping of $2v$ when the mass is given an initial displacement of 0.2 m and then released, with zero velocity.

Equation [14] gives

$$m\frac{d^2x}{dt^2} + c\frac{dx}{dt} + kx = 0$$

and hence the differential equation is

$$\frac{d^2x}{dt^2} + 2\frac{dx}{dt} + 9x = 0$$

This gives an auxiliary equation of

$$s^2 + 2s + 9 = 0$$

The roots of the quadratic equation are

$$s = \frac{-2 \pm \sqrt{4 - 36}}{2}$$

$$= -1 \pm j2.83$$

Hence the motion is underdamped and is described by an equation of the form given in equation [18] of

$$x = e^{-t}(C\cos 2.83t + D\sin 2.83t)$$

Since $x = 0.2$ at $t = 0$ then $C = 0.2$. The velocity is zero at $t = 0$ and since differentiating the above equation gives

$$\frac{dx}{dt} = e^{-t}(-2.83C\sin 2.83t + 2.83D\cos 2.83t)$$

$$- e^{-t}(C\cos 2.83t + D\sin 2.83t)$$

then $0 = 2.83D - C$. Thus $D = 0.0707$.

$$x = e^{-t}(0.2\cos 2.83t + 0.0707\sin 2.83t)$$

Review problems

5 An object of mass 1 kg is suspended from a spring with a stiffness of 4 N/m. What is the differential equation describing the motion and its solution when the mass is subject to a damping force of (a) $5v$, (b) $2v$?

6 An object of mass 1 kg is suspended from a spring with a stiffness of 9 N/m. What is the damping force per unit velocity for viscous damping which would give a critically damped motion of the mass?

7 An object of mass 1 kg is suspended from a spring with a stiffness of 1 N/m. What is the equation describing how the displacement varies with time if the mass is given an initial displacement of 1 m and released with zero velocity and the damping force per unit velocity is (a) 0, (b) 2 N per m/s, (c) 4 N per m/s?

8 An object of mass 0.5 kg is suspended from a spring with a stiffness of 4 N/m. The mass is subject to a damping force of 3 N per m/s. What is the equation describing how the displacement varies with time if the mass is given an initial displacement of 2 m with zero initial velocity?

9 An object of mass 1 kg is suspended from a spring of stiffness 100 N/m. The mass is subject to a viscous damping force of 100 N per m/s. What is the frequency of the damped oscillation?

7.3 Forced mechanical oscillations

Consider a mass suspended from a spring with the mass subject to a force which varies with time. We will suppose that this applied force varies with time according to $F\cos\omega_f t$, where F is the maximum value of the force and ω_f its angular frequency. The resulting oscillations of the mass are said to be *forced*, with ω_f the forcing frequency. If the mass is displaced some distance x from its equilibrium position then the spring will exert a restoring force of $-kx$, where k is the spring stiffness. Thus the resultant force acting on the mass at this displacement, if we assume that there is no damping force, is

$$\text{resultant force} = F\cos\omega_f t - kx$$

Hence, applying Newton's second law,

$$m\frac{d^2x}{dt^2} = F\cos\omega_f t - kx$$

This can be written in the form

$$m\frac{d^2x}{dt^2} + kx = F\cos\omega_f t \qquad [20]$$

This is a second-order, non-homogeneous, differential equation which can be solved in the way discussed in section 6.4.

Consider the solution of the above differential equation in terms of the complementary function x_c and the particular integral x_p, i.e.

$$x = x_c + x_p$$

The complementary function is the solution of the corresponding homogeneous equation. Thus, since this is

$$m\frac{d^2x}{dt^2} + kx = 0$$

the auxiliary equation is

$$ms^2 + ks = 0$$

and so the roots are $\pm j\sqrt{(k/m)}$. Writing $\omega_n = \sqrt{(k/m)}$ then we can write the solution in the form (see earlier in this chapter)

$$x_c = A\cos\omega_n t + B\sin\omega_n t$$

with A and B being constants and ω_n the angular frequency of the natural oscillations of the mass, i.e. the oscillations that would occur in the absence of any forcing force applied to the mass. For the particular integral we can try a solution of the form (see section 6.4)

$$x_p = C\cos\omega_f t + D\sin\omega_f t$$

Thus using this solution gives, for the differential equation,

$$m(-\omega_f^2 C\cos\omega_f t - \omega_f^2 D\sin\omega_f t) \\ + k(C\cos\omega_f t + D\sin\omega_f t) = F\cos\omega_f t$$

Thus, equating the cosine terms gives

$$-m\omega_f^2 C + kC = F$$

and so

$$C = \frac{F}{k - m\omega_f^2}$$

We can write $\omega_n = \sqrt{k/m}$ and thus

$$C = \frac{F}{m\left(\omega_n^2 - \omega_f^2\right)}$$

Equating the sine terms gives

$$-m\omega_f^2 D + kD = 0$$

Thus $D = 0$. Hence

$$x_p = \frac{F}{m\left(\omega_n^2 - \omega_f^2\right)} \cos \omega_f t$$

Thus the solution of the differential equation for the forced motion (equation [20]) is

$$x = x_c + x_p$$

$$= A \cos \omega_n t + B \sin \omega_n t + \frac{F}{m\left(\omega_n^2 - \omega_f^2\right)} \cos \omega_f t \qquad [21]$$

The resulting motion is thus the superposition of two oscillations, one with the natural angular frequency ω_n and the other with the forcing frequency ω_f. Figure 7.10 shows the resulting displacement when the difference between the natural and forcing frequencies is small. Beats occur.

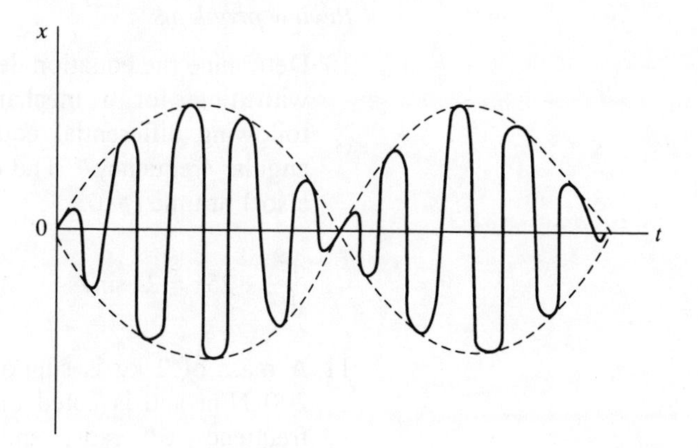

Fig. 7.10 When the difference between the forcing and natural frequencies is small

As $\omega_f \to \omega_n$ then the amplitude of the particular integral term tends to infinity. The consequence of this is that the amplitude of the oscillations tend to infinity, as indicated in figure 7.11. When $\omega_f = \omega_n$ then the amplitude is infinite and the system is said to be in *resonance*.

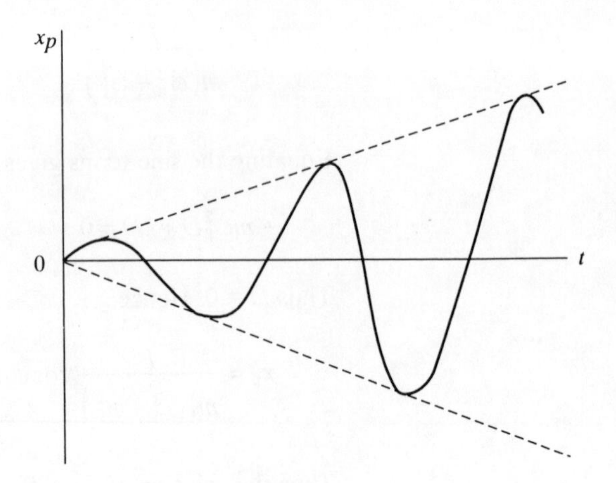

Fig. 7.11 Resonance

Example

Determine the angular natural frequency and the forcing frequency for a mechanical system corresponding to the differential equation

$$\frac{d^2x}{dt^2} + 25x = 4 \sin 3t$$

The differential equation can be compared with equation [20]. Thus the forcing angular frequency is 3 rad/s and the natural angular frequency is $\sqrt{(k/m)} = \sqrt{25} = 5$ rad/s.

Review problems

10 Determine the equation describing how the displacement varies with time for a mechanical system corresponding to the following differential equation. What will be the resonance angular frequency? The displacement x is 1 and the velocity also 1 at time $t = 0$.

$$\frac{d^2x}{dt^2} + 25x = 24 \sin t$$

11 A mass of 2 kg is supported by a vertical spring of stiffness 200 N/m and is acted on by an alternating force of angular frequency 30 rad/s and amplitude 40 N. Determine the equation describing how the displacement varies with time. At time $t = 0$ the displacement x is 2.5 mm and the velocity zero.

7.3.1 Damped forced oscillations

If there is viscous damping then the mass on the spring will be subject to a damping force of cv, where v is the velocity. The velocity $v = \mathrm{d}x/\mathrm{d}t$. Thus, with damping, equation [20] becomes modified to become

$$m\frac{\mathrm{d}^2x}{\mathrm{d}t^2} + c\frac{\mathrm{d}x}{\mathrm{d}t} + kx = F\cos\omega_\mathrm{f}t \qquad [22]$$

The solution of this equation can be obtained in the same way as given in section 6.4 for the undamped forced oscillation. For the corresponding homogeneous equation we thus have the auxiliary equation

$$ms^2 + cs + k = 0$$

This has the roots

$$s = \frac{-c \pm \sqrt{c^2 - 4mk}}{2m}$$

$$= -\frac{c}{2m} \pm \mathrm{j}\frac{\sqrt{c^2 - 4mk}}{2m}$$

This can be written as

$$s = -\mu \pm \mathrm{j}\omega$$

and so gives (see equation [18]) the solution

$$x_\mathrm{c} = \mathrm{e}^{-\mu t}(A\cos\omega_\mathrm{n}t + B\sin\omega_\mathrm{n}t)$$

For the particular integral the solution is of the form

$$x_\mathrm{p} = C\cos\omega_\mathrm{f}t + D\sin\omega_\mathrm{f}t$$

Thus, using this solution gives

$$m(-\omega_\mathrm{f}^2C\cos\omega_\mathrm{f}t - \omega_\mathrm{f}^2D\sin\omega_\mathrm{f}t) + c(-\omega_\mathrm{f}C\sin\omega_\mathrm{f}t + \omega_\mathrm{f}D\cos\omega_\mathrm{f}t) + k(C\cos\omega_\mathrm{f}t + D\sin\omega_\mathrm{f}t) = F\cos\omega_\mathrm{f}t$$

Equating the cosine terms gives

$$-m\omega_\mathrm{f}^2C + c\omega_\mathrm{f}D + kC = F$$

Equating the sine terms gives

$$-m\omega_\mathrm{f}^2D - c\omega_\mathrm{f}C + kD = 0$$

These two simultaneous equations can be solved to give

$$C = \frac{F\left(k - m\omega_f^2\right)}{\left(k - m\omega_f^2\right)^2 + \omega_f^2 c^2}$$

$$D = \frac{F\omega_f c}{\left(k - m\omega_f^2\right)^2 + \omega_f^2 c^2}$$

Using $\omega_n = \sqrt{(k/m)}$ gives

$$C = \frac{Fm\left(\omega_n^2 - \omega_f^2\right)}{m^2\left(\omega_n^2 - \omega_f^2\right)^2 + \omega_f^2 c^2} \qquad [23]$$

$$D = \frac{F\omega_f c}{m^2\left(\omega_n^2 - \omega_f^2\right)^2 + \omega_f^2 c^2} \qquad [24]$$

Hence the general solution for the forced damped oscillation is

$$x = x_c + x_p$$

$$= e^{-\mu t}(A \cos \omega_n t + B \sin \omega_n t) + C \cos \omega_f t + D \sin \omega_f t \qquad [25]$$

The complementary function part of the solution will tend to zero as time increases, because of the exponential term. Thus the complementary function part of the solution is referred to as the *transient part of the response*. The steady state oscillations which will eventually prevail are given by the particular integral part of the solution and thus

$$\text{steady state } x = C \cos \omega_f t + D \sin \omega_f t \qquad [26]$$

With undamped forced oscillations, the amplitude of the oscillations becomes infinite when the forcing frequency becomes equal to the natural frequency. With damping the amplitude will always be finite but will have a maximum value at some frequency.

Example

Determine the steady state oscillations that will occur for a mechanical system corresponding to the differential equation

$$\frac{d^2x}{dt^2} + 3\frac{dx}{dt} + 2x = 12 \cos t$$

The forcing angular frequency is 1 and the natural angular frequency is √(2/1). The steady state solution is given by equation [26], with the terms C and D being given by equations [23] and [24]. Thus

$$C = \frac{Fm\left(\omega_n^2 - \omega_f^2\right)}{m^2\left(\omega_n^2 - \omega_f^2\right)^2 + \omega_f^2 c^2}$$

$$= \frac{12 \times 1(2-1)}{1(2-1)^2 + 1 \times 9} = 1.2$$

$$D = \frac{F\omega_f c}{m^2\left(\omega_n^2 - \omega_f^2\right)^2 + \omega_f^2 c^2}$$

$$= \frac{12 \times 1 \times 3}{1(2-1)^2 + 1 \times 9} = 3.6$$

Hence the steady state response is

$$x = 1.2\cos t + 3.6\sin t$$

Review problems

12 Determine the steady state oscillations of the mechanical system corresponding to the differential equation

$$2\frac{d^2x}{dt^2} + 6\frac{dx}{dt} + 4x = 20\sin t$$

13 Determine the steady state oscillations of the mechanical system corresponding to the differential equation

$$\frac{d^2x}{dt^2} + 6\frac{dx}{dt} + 10x = 30\cos 2t$$

Further problems

14 An object of mass 0.5 kg is attached to the end of a vertically suspended spring. The spring obeys Hooke's law and is stretched 0.2 m by a force of 50 N. The mass is pulled down a distance of 0.1 m and then released. Determine the differential equation for the motion and the position of the object after a time of 0.1 s.

15 An object of mass of 0.5 kg is suspended from a vertical spring. If the spring obeys Hooke's law and has a stiffness of 18 N/m, what is the natural period of oscillation of the object? If, when the object is stretching the spring by 0.042 m, it is given a velocity of 0.33 m/s in the direction of x increasing at $t = 0$, what will be the amplitude of the oscillations?

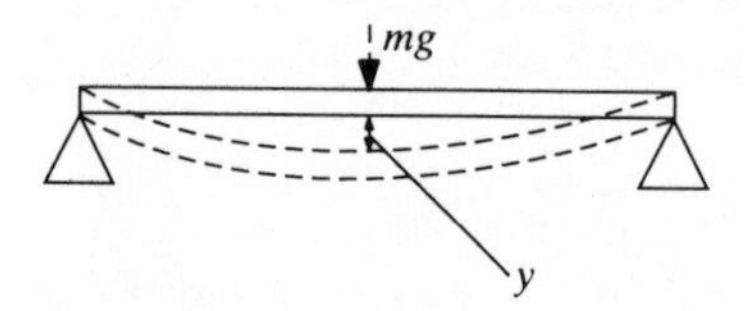

Fig. 7.12 Problem 16

16 Derive an equation for the periodic time of natural oscillations of a simply supported beam, as illustrated in figure 7.12, when subject to a single concentrated load of mg applied to its mid-point. The deflection y of the mid-point of the beam is related to the force F at its centre point required to bend it to that deflection by $F = 48EIy/L^3$.

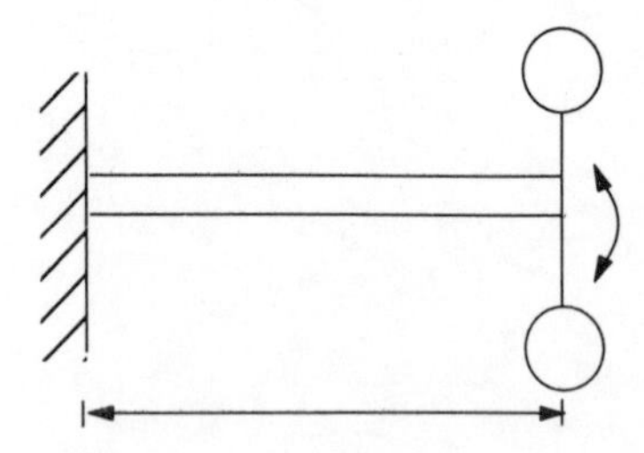

Fig. 7.13 Problem 17

17 Derive an equation for the natural torsional oscillations of a system consisting of a rotor attached to the end of a shaft, as illustrated in figure 7.13. The rotor has a moment of inertia I about the axis of the shaft, the rod a length L, second moment of area J and modulus of rigidity G, with the torque T needed to twist the shaft through an angle θ being given by the equation $T = GJ\theta/L$.

18 A mass of 1 kg hangs from a vertical spring of stiffness 9 N/m. There is a damping force of $1v$ opposing the motion of the mass. Determine the equation for the variation of the displacement with time for (a) when the mass is given an initial displacement of 0.2 m and released with zero initial velocity, (b) when the mass is given an initial displacement of 0.2 m and released with an initial velocity of −0.3 m/s.

19 Is the frequency of a damped oscillating system (a) the same as the natural frequency, (b) dependent on the initial displacement and velocity conditions?

20 A cannon has a barrel of mass 500 kg. After firing a shell the barrel recoils. It is returned to its original position by means of two parallel springs. Determine the stiffness required of the springs if the barrel returns to its original position in the minimum time without any oscillation.

21 A ship rolls in a calm sea with an angle of roll θ which varies with time according to the differential equation

$$\frac{d^2\theta}{dt^2} + 2\alpha^2\frac{d\theta}{dt} + \beta^2\theta = 0$$

Determine the solution of the differential equation if α^2 can be ignored in comparison with β^2.

22 An object of mass 6 kg is suspended from a spring of stiffness 1000 N/m. If the mass is subject to a viscous resistance of 36 N per m/s, what will the relation between the displacement and time when the object is pulled down a distance of 50 mm and released with zero initial velocity?

23 Determine how the displacement varies with time for a

mechanical system corresponding to the differential equation

$$\frac{d^2x}{dt^2} + x = 3\cos 2t$$

24 Determine how the displacement varies with time for a mechanical system corresponding to the differential equation

$$\frac{d^2x}{dt^2} + 9x = 16\cos 5t$$

The displacement x is 0 and the velocity 0 at time $t = 0$.

25 A mass of 1 kg is suspended from a spring of stiffness 9 N/m. If the system is subject to a periodic force of amplitude 15 N and angular frequency 4 rad/s, how will the displacement of the mass vary with time?

26 Determine how the displacement varies with time for the mechanical system corresponding to the differential equation

$$\frac{d^2x}{dt^2} + 2\frac{dy}{dx} + 2x = 20\cos 2t$$

if the displacement x is zero and the velocity zero at time $t = 0$.

27 What will be the natural angular frequency and the forcing angular frequency for a mechanical system corresponding to the differential equation

$$\frac{d^2x}{dt^2} + 5\frac{dx}{dt} + 4x = 10\cos 3t$$

28 A mass of 1 kg is suspended from a spring of stiffness 2 N/m and subject to a damping force of $0.5v$, where v is the velocity in m/s. Determine how the displacement will vary with time when the system is subject to a force of 5 cos t.

29 Determine how the displacement will vary with time for a mechanical system corresponding to the following differential equation

$$\frac{d^2x}{dt^2} + 6\frac{dx}{dt} + 10x = 30\cos 2t$$

30 Determine how the displacement will vary with time for a mechanical system corresponding to the following differential equation

$$\frac{d^2x}{dt^2} + 2\frac{dx}{dt} + 10x = 120\sin 4t$$

given that the displacement x and the velocity are both zero at time $t = 0$.

8 Second-order differential equations: electric circuits

8.1 *RLC* series circuit

This chapter can be considered to be an extension of chapter 3. In that chapter the circuits discussed involved just resistance and capacitance or resistance and inductance. Such circuits gave first-order differential equations. In this chapter circuits involving resistance, inductance and capacitance are considered. Such circuits give second-order differential equations.

Consider an electrical circuit having a resistor, inductor and capacitor in series, as in figure 8.1. Applying Kirchhoff's voltage law, then when the switch is closed and the voltage V applied, we must have

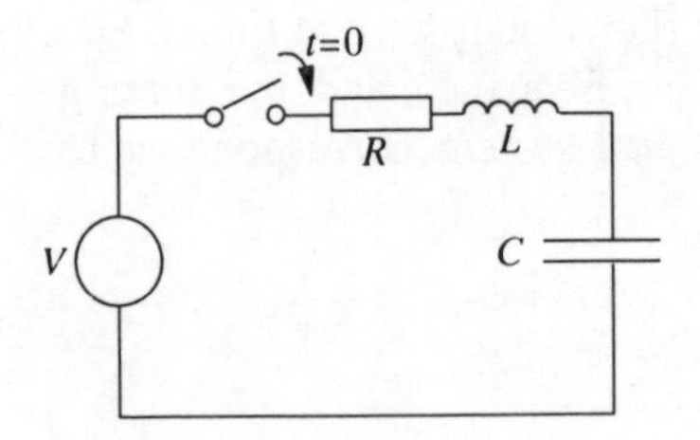

Fig. 8.1 *RLC* series circuit

$$v_R + v_L + v_C = V$$

where V is the input voltage to the circuit, v_R is the potential difference across the resistor, v_L the potential difference across the inductor and v_C the potential difference across the capacitor. We can write $v_R = Ri$ (equation [1], chapter 3), $v_L = L\,\mathrm{d}i/\mathrm{d}t$ (equation [6], chapter 3). Thus

$$Ri + L\frac{\mathrm{d}i}{\mathrm{d}t} + v_C = V$$

But $i = C\,\mathrm{d}v_C/\mathrm{d}t$ (equation [3], chapter 3). Hence

$$LC\frac{\mathrm{d}^2 v_C}{\mathrm{d}t^2} + RC\frac{\mathrm{d}v_C}{\mathrm{d}t} + v_C = V \qquad [1]$$

For convenience, in the discussion that follows the subscript C used to indicate that the potential difference refers to that across the capacitor will be dropped.

This second-order differential equation can be solved by the method given in section 6.4 for non-homogeneous equations. Thus consider the solution to be the sum of the complementary function

v_c and the particular integral v_p. The complementary function is the solution of the corresponding homogeneous differential equation, namely

$$LC\frac{d^2v_c}{dt^2} + RC\frac{dv_c}{dt} + v_c = 0$$

This represents what is termed the *transient* part of the voltage since it will eventually die away, leaving just that element of the solution given by the particular integral. The differential equation in fact describes the circuit in the absence of any voltage, or current source.

Trying a solution of the form $v_c = A\,e^{st}$ gives an auxiliary equation

$$LCs^2 + RCs + 1 = 0$$

The roots of this quadratic equation are

$$s = \frac{-RC \pm \sqrt{(RC)^2 - 4LC}}{2LC}$$

$$= -\frac{R}{2L} \pm \sqrt{\left(\frac{R}{2L}\right)^2 - \frac{1}{LC}} \qquad [2]$$

There are three possible forms of roots that can occur (see section 6.3.1):

1 $(R/2L)^2 > (1/LC)$
For this condition there will be two real distinct roots and the solution will be of the form

$$v_c = A\,e^{s_1 t} + B\,e^{s_2 t} \qquad [3]$$

The circuit is said to be *overdamped* with the transient part of the voltage just slowly decaying with time to zero.

2 $(R/2L)^2 = (1/LC)$
For this condition the two roots are real and equal. The solution is of the form

$$v_c = (At + B)\,e^{st} \qquad [4]$$

The circuit is said to be *critically damped*. The transient part of the voltage decays to leave just the steady state value in the minimum amount of time without any oscillations occurring.

3 $(R/2L)^2 < (1/LC)$

For this condition the two roots are complex and of the form $-a \pm j\omega$, with a being $(R/2L)$ and ω as $\sqrt{[(1/LC) - (R/2L)^2]}$. The solution is of the form

$$v_c = e^{-at}(A \cos \omega t + B \sin \omega t) \qquad [5]$$

The circuit is said to be *underdamped* and the transient part of the voltage oscillates about the steady state value before eventually dying away. In the absence of resistance then a would be zero and then the solution would be

$$v_c = A \cos \omega_n t + B \sin \omega_n t$$

The oscillations would then continue indefinitely. The angular frequency in this case ω_n is termed the natural angular frequency and is $\omega_n = \sqrt{(1/LC)}$.

The particular integral for equation [1] will be, for a constant voltage input of V, of the form $v_p = k$, where k is a constant. Thus substituting this into equation [1] gives

$$0 + 0 + k = V$$

Thus the particular integral is $v_p = V$.

Hence the general solution for the differential equation [1] is, depending on the form of the roots of the corresponding homogeneous differential equation,

$$v = A e^{s_1 t} + B e^{s_2 t} + V \qquad [6]$$

$$v = (At + B) e^{st} + V \qquad [7]$$

$$v = e^{-at}(A \cos \omega t + B \sin \omega t) + V \qquad [8]$$

Figure 8.2 shows a graph of how the voltage across the capacitor varies with time for the above three situations.

Equation [8] is often written in a different form. If we consider a right-angled triangle of sides A, B and V_m, as in figure 8.3, then we have $A/V_m = \cos \alpha$ and $B/V_m = \sin \alpha$. Thus equation [8] becomes

$$v = e^{-at}(V_m \cos \omega t \cos \alpha + V_m \sin \omega t \sin \alpha) + V$$

Since $\cos(a - b) = \cos a \cos b + \sin a \sin b$ then we can write

$$v = V_m e^{-at} \cos(\omega t - \alpha) + V \qquad [9]$$

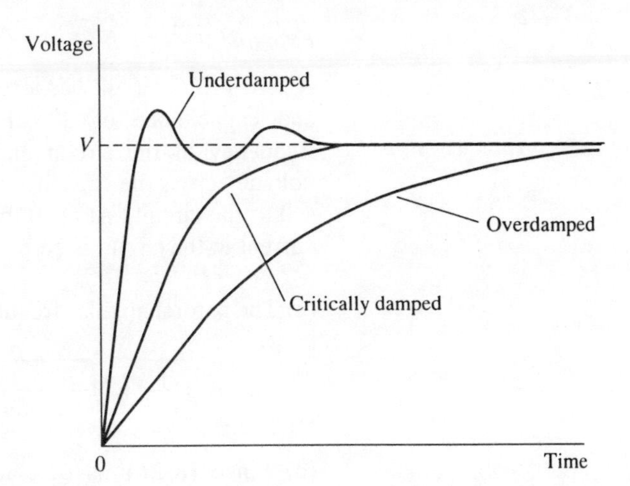

Fig. 8.2 Responses of the *RLC* series circuit

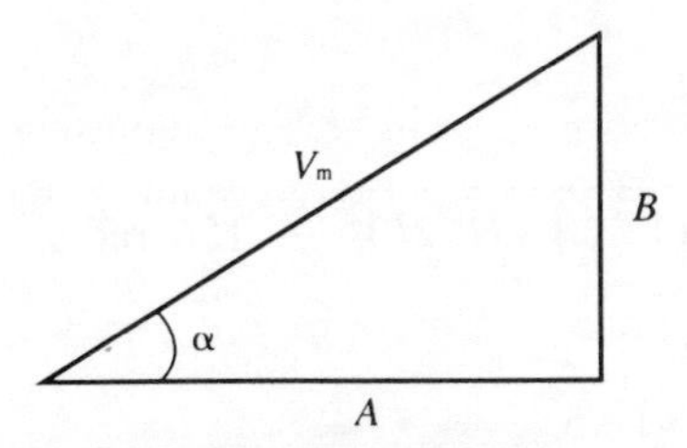

Fig. 8.3 Phase angle α

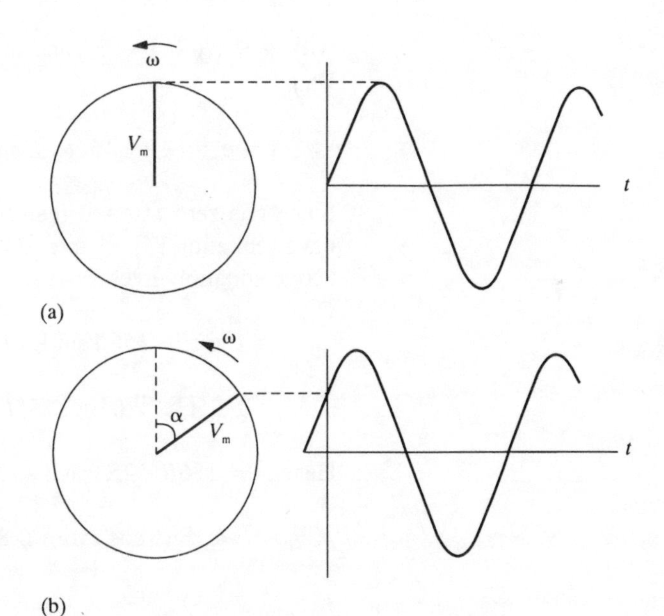

Fig. 8.4 (a) $v = V_m \cos \omega t$, (b) $v = V_m \cos(\omega t - \alpha)$

$V_m \cos \omega t$ can be considered to describe a cosine wave generated by a line of length V_m rotating with an angular velocity ω, as in figure 8.4(a). V_m is thus the maximum value of the voltage. The $V_m \cos(\omega t - \alpha)$ can be considered to describe the wave generated by a line of length V_m rotating with an angular velocity ω but starting with a phase lag of α, as in figure 8.4(b). The expression $V_m e^{-at} \cos(\omega t - \alpha)$ thus describes a wave with an amplitude $V_m e^{-at}$, such an amplitude diminishing with time.

Example

A series *RLC* circuit has a resistance of 100 Ω, inductance of 2 H and capacitance of 20 μF. What is (a) the natural angular frequency of the circuit and (b) the variation with time of the voltage across the capacitor when there is an input of 5 V at time t = 0 to the circuit? At $t = 0$ the voltage across the capacitor and the current in the circuit is zero.

(a) The natural angular frequency is given by

$$\omega_n = \sqrt{\frac{1}{LC}} = \sqrt{\frac{1}{2 \times 20 \times 10^{-6}}} = 158 \text{ rad/s}$$

(b) Since $(R/2L)^2$ is 625 and $(1/LC)$ is 25 000, then we have $(R/2L)^2 < (1/LC)$ and so the circuit is underdamped. The solution is thus of the form given by equation [8].

$$v = e^{-at}(A \cos \omega t + B \sin \omega t) + V$$

with $a = (R/2L) = 25$ and $\omega = \sqrt{(1/LC) - (R/2L)^2} = 156$ rad/s. Thus

$$v = e^{-25t}(A \cos 156t + B \sin 156t) + 5$$

Since v is zero at $t = 0$ then $0 = A + 5$ and so $A = -5$. Since we have (equation [3], chapter 3) $i = C\,dv/dt$ then differentiating the above equation gives

$$i = Ce^{-25t}(-156A \sin 156t + 156B \cos 156t)$$

$$-25Ce^{-25t}(A \cos 156t + B \sin 156t)$$

Hence $0 = 156B - 25A$ and so $B = -0.80$. Thus, in volts,

$$v = -e^{-25t}(5 \cos 156t + 0.80 \sin 156t) + 5$$

Review problems

1 A 1 F capacitor which has been charged to 10 V is discharged at time $t = 0$ through a series circuit of resistance 2 Ω and inductance 1 H. If the initial current is 0, derive an equation describing how the voltage across the capacitor varies with time.

2 A 0.04 F capacitor is in series with a 1 H inductor and a 5 Ω resistor. Is the circuit over-, critically or underdamped? What would the resistance need to be for the circuit to be critically damped?

3 A circuit consists of a 0.05 μF capacitor in series with an inductance of 0.5 H and a resistance of 4 kΩ. How will the voltage across the capacitor vary with time when a 1 V supply is connected to the circuit at time $t = 0$? The voltage across the capacitor and the circuit current are zero at the time $t = 0$.

4 For the circuit shown in figure 8.5, determine the conditions for the circuit to be over-, critically and underdamped. (Hint: apply Kirchhoff's voltage law to give one equation involving the voltages across each part of the circuit and then obtain another equation resulting from the voltage across the resistor being equal to that across the inductor.)

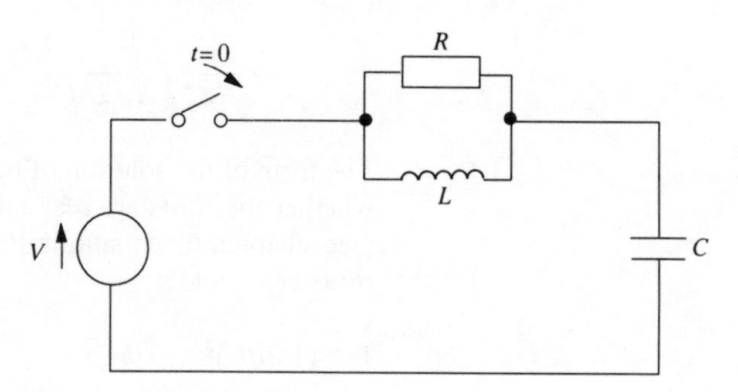

Fig. 8.5 Problem 4

8.2 *RLC* parallel circuit

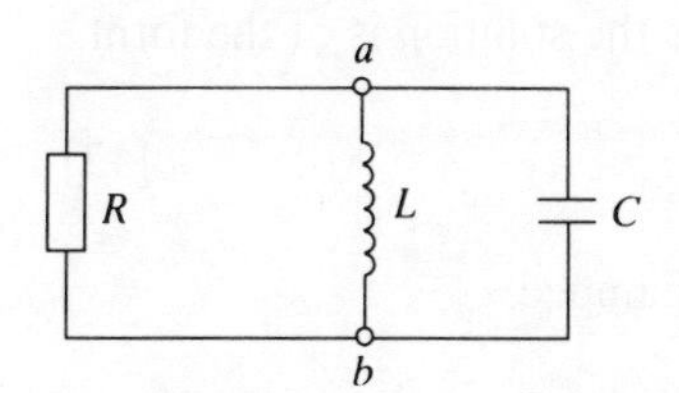

Fig. 8.6 Parallel *RLC* circuit

Consider the circuit shown in figure 8.6 which has a resistor, an inductor and a capacitor connected in parallel. If we apply Kirchhoff's current law to node a then we must have the algebraic sum of the currents through the three components equal to zero. If v is the potential difference between nodes a and b then the current through the resistor is v/R (equation [1], chapter 3), through the inductor is $(1/L)\int v\,\mathrm{d}t$ (equation [7], chapter 3), and through the capacitor is $C\,\mathrm{d}v/\mathrm{d}t$ (equation [3], chapter 3). Thus

$$\frac{v}{R} + \frac{1}{L}\int_0^t v\,\mathrm{d}t + C\frac{\mathrm{d}v}{\mathrm{d}t} = 0$$

If the equation is differentiated with respect to t then we obtain

$$\frac{1}{R}\frac{\mathrm{d}v}{\mathrm{d}t} + \frac{v}{L} + C\frac{\mathrm{d}^2 v}{\mathrm{d}t^2} = 0$$

This can be rearranged to give

$$\frac{\mathrm{d}^2 v}{\mathrm{d}t^2} + \frac{1}{RC}\frac{\mathrm{d}v}{\mathrm{d}t} + \frac{1}{LC}v = 0 \qquad [10]$$

If we assume a solution of the form $v = A\,e^{st}$ then we obtain the auxiliary equation

$$s^2 + \frac{1}{RC}s + \frac{1}{LC} = 0$$

The roots of this quadratic equation are thus

$$s = \frac{-\frac{1}{RC} \pm \sqrt{\left(\frac{1}{RC}\right)^2 - 4\left(\frac{1}{LC}\right)}}{2}$$

$$= -\frac{1}{2RC} \pm \sqrt{\left(\frac{1}{2RC}\right)^2 - \left(\frac{1}{LC}\right)} \qquad [11]$$

The form of the solution of the differential equation will depend on whether the roots are real and distinct, real and equal, or complex (see chapter 6, equations [9], [10] and [11]). Thus, if the two roots are s_1 and s_2,

1 $(1/2RC)^2 > (1/LC)$
The roots are real and distinct and so the solution is of the form

$$v = A\,e^{s_1 t} + B\,e^{s_2 t} \qquad [12]$$

The circuit is said to be overdamped.

2 $(1/2RC)^2 = (1/LC)$
The roots are real and equal. Thus the solution is of the form

$$v = (At + B)\,e^{st} \qquad [13]$$

The circuit is said to be critically damped.

3 $(1/2RC)^2 < (1/LC)$
The roots are real and complex, being of the form $-a \pm j\omega$, with a being $1/2RC$ and ω equal to $\sqrt{[(1/2RC)^2 - (1/LC)]}$. The solution is thus

$$v = e^{-at}(A\cos\omega t + B\sin\omega t) \qquad [14]$$

In the absence of damping, i.e. $R = 0$, then the angular frequency is termed the natural or free angular frequency ω_n and is $\sqrt{(1/LC)}$.

Example

A circuit consists of three components in parallel, a resistance of 10 Ω, a 0.5 H inductance and a 0.05 F capacitor. At a time $t = 0$ the potential difference across the capacitor is 10 V and the current through the inductor is 2 A. Derive the equation relating the voltage across the components with time.

For this circuit we have $(1/2RC)^2 = 1$ and $(1/LC) = 40$. Thus the circuit is underdamped. The solution is thus of the form given by equation [14], i.e.

$$v = e^{-at}(A\cos\omega t + B\sin\omega t)$$

and so

$$v = e^{-t}(A\cos 6.24t + B\sin 6.24t)$$

At $t = 0$ we have $v = 10$ V. Thus $A = 10$. Also, at $t = 0$, we have the current through the inductor as 2 A. Because the voltage across the components is 10 V at this time then the current through the resistor must be $10/10 = 1$ A. Thus, since the sum of the currents at a node in the circuit must be zero, we must have the current through the capacitor as -3 A. Thus, since the current through the capacitor is given by $C\,\mathrm{d}v/\mathrm{d}t$ we must have $\mathrm{d}v/\mathrm{d}t = -60$ V/s at time $t = 0$. Differentiating the above equation gives

$$\frac{\mathrm{d}v}{\mathrm{d}t} = e^{-t}(-6.24A\sin 6.24t + 6.24B\cos 6.24t) - e^{-t}(A\cos 6.24t + B\sin 6.24t)$$

Hence

$$-60 = 6.24B - A$$

Thus $B = -8$. Hence the solution is

$$v = e^{-t}(10\cos 6.24t - 8\sin 6.24t)$$

Review problems

5 A circuit consists of three parallel components, a 20 kΩ resistance, a 0.8 H inductance and a 0.125 μF capacitance. Determine how the voltage across the components will vary with time if initially the voltage is zero and the current through the inductor is -25 mA.

6 A circuit consists of three parallel components, a resistor, a 0.01 H inductance and a 1 μF capacitance. What should the value of the resistance be if the circuit is to be critically damped?

8.2.1 Parallel *RLC* circuit with current input

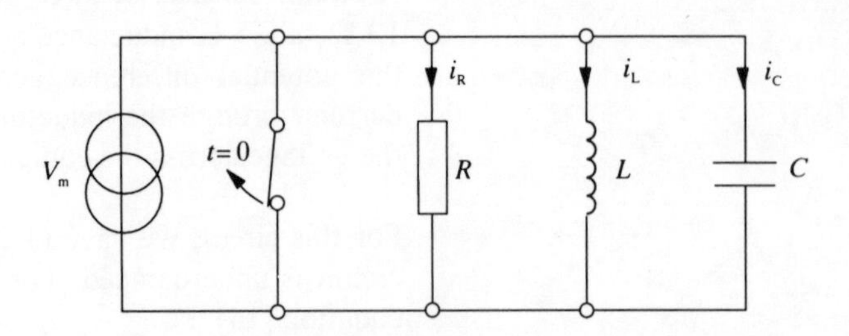

Fig. 8.7 Parallel *RLC* circuit

Consider the parallel *RLC* circuit when a steady current is applied to it at time $t = 0$. Figure 8.7 shows the circuit. At a time of $t = 0$, when the switch is opened, then Kirchhoff's current law gives the algebraic sum of the currents through the resistor, the inductor and the capacitor being equal to I, the applied current. Thus

$$i_R + i_L + i_C = I$$

But $i_R = v/R$ and $i_C = C\ dv/dt$, where v is the potential difference across the three components. Since we also have $v = L\ di_L/dt$ then we can write $i_R = (L/R)\ di_L/dt$ and $i_C = LC\ d^2i_L/dt^2$. Thus

$$\frac{L}{R}\frac{di_L}{dt} + i_L + LC\frac{d^2i_L}{dt^2} = I$$

This can be rearranged to give

$$\frac{d^2i_L}{dt^2} + \frac{1}{RC}\frac{di_L}{dt} + \frac{1}{LC}i_L = \frac{I}{LC} \qquad [15]$$

This is a non-homogeneous second-order differential equation.

The differential equation can be considered to have a solution which is the sum of the complementary function and the particular integral. The complementary function is the solution of the corresponding homogeneous equation, namely

$$\frac{d^2i_L}{dt^2} + \frac{1}{RC}\frac{di_L}{dt} + \frac{1}{LC}i_L = 0$$

This is of the same form as equation [10] and thus has similar solutions. The particular integral solution that can be tried is $i_L = k$, where k is a constant. Thus, using this value in the differential equation gives

$$0 + 0 + \frac{k}{LC} = \frac{I}{LC}$$

Thus the particular integral solution is $i_L = I$. Hence the solution to the non-homogeneous differential equation is:

1 $(1/2RC)^2 > (1/LC)$
The roots are real and distinct and so the solution is of the form

$$i_L = A\,e^{s_1 t} + B\,e^{s_2 t} + I \qquad [16]$$

The circuit is said to be overdamped.

2 $(1/2RC)^2 = (1/LC)$
The roots are real and equal. Thus the solution is of the form

$$i_L = (At + B)\,e^{st} + I \qquad [17]$$

The circuit is said to be critically damped.

3 $(1/2RC)^2 < (1/LC)$
The roots are real and complex, being of the form $-a \pm j\omega$, with a being $1/2RC$ and ω equal to $\sqrt{[(1/2RC)^2 - (1/LC)]}$. The solution is thus

$$i_L = e^{-at}(A\cos\omega t + B\sin\omega t) + I \qquad [18]$$

The circuit is said to be underdamped.

Example

A circuit consists of three parallel elements, a 100 Ω resistance, a 25 mH inductor and a 0.2 μF capacitor. How will the current through the inductor vary with time, and how will the potential difference across the components vary with time, when a d.c. current source of 20 mA is applied to it at time $t = 0$, the current in the circuit and the voltage across the components prior to that time being zero?

$(1/2RC)^2$ has the value 6.25×10^8 and $(1/LC) = 2 \times 10^8$. The circuit is thus overdamped. Hence the solution is of the form given by equation [16], namely

$$i_L = A\,e^{s_1 t} + B\,e^{s_2 t} + I$$

The roots of the auxiliary equation are given by equation [11] as

$$s = -\frac{1}{2RC} \pm \sqrt{\left(\frac{1}{2RC}\right)^2 - \left(\frac{1}{LC}\right)}$$

$$= -25\,000 \pm \sqrt{6.25 \times 10^8 - 2 \times 10^8}$$

$$= -25\,000 \pm \sqrt{6.25 \times 10^8 - 2 \times 10^8}$$

$$= -25\,000 \pm 20\,616$$

Thus the roots are $-45\,616$ and -4384 and so

$$i_L = A\,e^{-45\,616t} + B\,e^{-4384t} + I$$

The current I is 0.020 A. Since $i_L = 0$ at $t = 0$ then

$$0 = A + B + 0.020$$

Since $v = L\,di_L/dt$ then

$$v = L(-45\,616A\,e^{-45\,616t} - 4384B\,e^{-4384t})$$

With $v = 0$ at $t = 0$ then

$$45\,616A + 4384B = 0$$

The two simultaneous equations in A and B can be solved to give $A = 0.042$ and $B = -0.022$. Thus

$$i_L = 0.042\,e^{-45\,616t} - 0.022\,e^{-4384t} + 0.02 \text{ A}$$

The potential difference across the inductor is $L\,di_L/dt$ and thus is

$$v = 0.025(-0.042 \times 45\,616\,e^{-45\,616t} + 0.022 \times 4384\,e^{-4384t})$$

$$= 0.025(-1916\,e^{-45\,616t} + 96.45\,e^{-4384t}) \text{ V}$$

Review problems

7 A circuit consists of three parallel components, a resistance of 50 Ω, an inductance of 0.02 H, and a capacitance of 2.5 μF. How will the current through the capacitor vary with time if at time $t = 0$ a current of 2 A is applied to the circuit? Prior to that time there is no current in the circuit and no potential difference across the components.

8 A circuit consists of a 0.2 F capacitor in parallel with an inductor of resistance 2 Ω and inductance 250 mH. Determine how the current through the inductor will vary with time when, at time $t = 0$, a current source of 10 A is applied to the circuit. Prior to that time there is no current in the circuit and no potential difference across the capacitor or the inductor.

9 A circuit consists of three components in parallel, a resistance

of 1 Ω, an inductance of 2 H and a capacitance of 0.5 F. Determine how the current through the inductor varies with time when a current source of 1 A is connected to it at time $t = 0$. Prior to this time there is no current in the circuit and no potential difference across the capacitor.

Further problems

10 A 0.05 F capacitor which has been charged to 10 V is discharged at time $t = 0$ through a resistance of 3 Ω and an inductance of 1 H. If the initial current is 0, derive an equation describing how the voltage across the capacitor will vary with time.

11 A series *RLC* circuit has a resistance of 2 kΩ, a capacitance of 2 μF and an inductance of 5 mH. Is the circuit over-, critically or underdamped?

12 For the circuit shown in figure 8.8, determine the condition that the circuit will be critically damped.

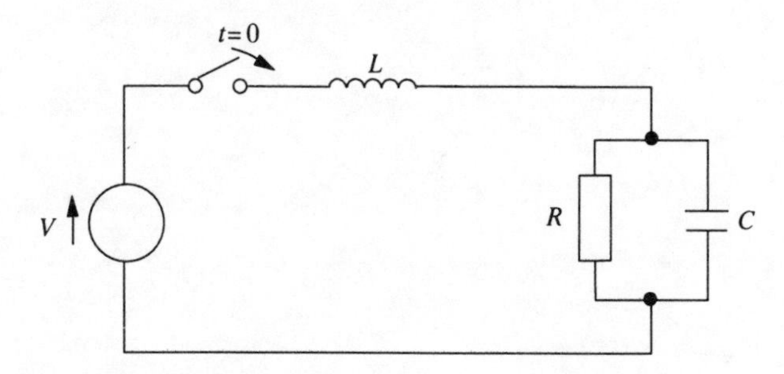

Fig. 8.8 Problem 12

13 A series *RLC* circuit has a resistance of 12 Ω, an inductance of 2 H and a capacitance of 0.02 F. At time $t = 0$ the voltage across the capacitor and the circuit current are zero. Determine how the voltage across the capacitor will vary with time when a voltage of 0.4 V is connected to it at $t = 0$.

14 Determine the condition for critical damping for the circuit shown in figure 8.9.

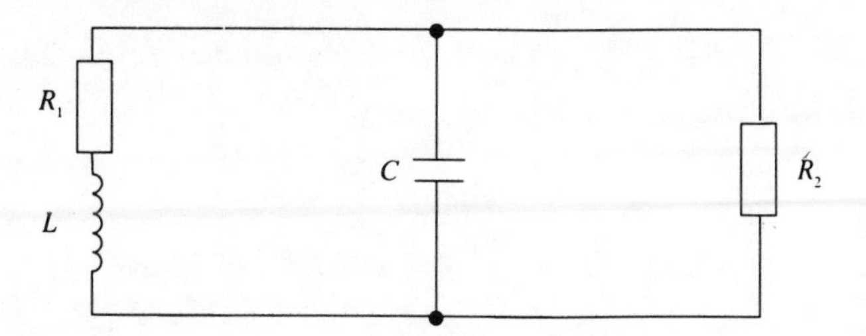

Fig. 8.9 Problem 14

15 A circuit consists of a 0.1 F capacitor in parallel with an inductor of resistance 6 Ω and inductance 1 H. Determine

how the potential difference across the capacitor will vary with time when a current source of 4 A is connected to the circuit at time $t = 0$. Prior to that time there was no current in the circuit and no voltage across the capacitor.

16 A circuit consists of three components in series, a resistance of 80 Ω, an inductance of 20 H and a capacitance of 10 mF. Determine how the current in the circuit varies with time when a voltage of 100 V is applied to it at time $t = 0$. Prior to that time there was no current in the circuit and no voltage across the capacitor.

9 Simultaneous differential equations

9.1 Simultaneous linear equations

Consider a particular situation where the velocity dy/dt of some body in the y direction is given by

$$\frac{dy}{dt} = 5 - 10t$$

and the velocity in the x direction, i.e. dx/dt, is given by

$$\frac{dx}{dt} = 4$$

These two differential equations could be describing the motion of a projectile, with the initial velocity component in the y, or vertical, direction being 5 m/s (the acceleration due to gravity has been taken as 10 m/s^2) and that in x, or horizontal, direction 4 m/s. Both differential equations have to be solved if the way in which the position of the body varies with time is to be determined. Thus solving each equation, for example by separation of variables, gives

$$y = 5t - 5t^2 + A$$

$$x = 4t + B$$

If y and x are both 0 at $t = 0$ then A and B are both 0. Thus

$$y = 5t - 5t^2$$

$$x = 4t$$

Thus when $t = 0.1$ then $y = 0.45$ and $x = 0.4$. With $t = 0.2$ then we have $y = 0.80$ and $x = 0.8$. By obtaining such values we can plot a graph of y against x, as in figure 9.1. This graph has an equation

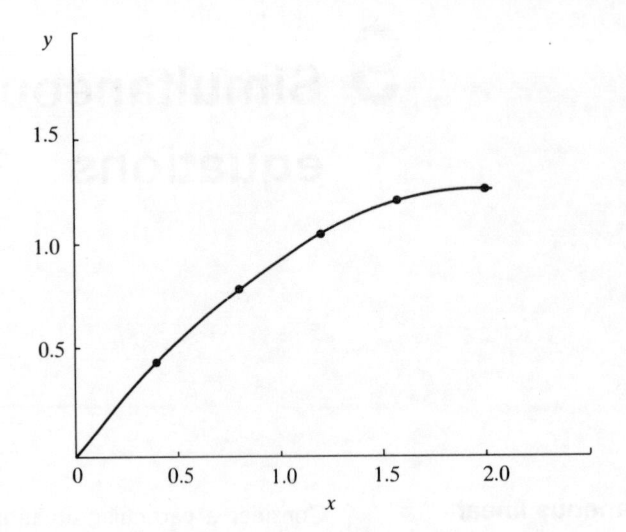

Fig. 9.1 The projectile motion

which we could obtain by solving the pair of simultaneous equations, e.g. by substituting $t = x/4$ in the equation for y. Thus

$$y = 5\left(\frac{x}{4} - \frac{x^2}{16}\right)$$

The above example involved the solution of the two simultaneous differential equations by analytical methods; they could, however, each have been solved by numerical methods (see chapter 5).

Now consider a slightly more complex example. In a chemical reaction involving two interacting substances, if x is the amount of one substance and y the amount of the other then we might, for example, have the two following differential equations describing how the amounts of the two substances vary with time:

$$\frac{dx}{dt} = -0.1x + 0.5y \qquad [1]$$

$$\frac{dy}{dt} = 0.1x - 0.5y \qquad [2]$$

With such equations, suppose we differentiate the first equation.

$$\frac{d^2x}{dt^2} = -0.1\frac{dx}{dt} + 0.5\frac{dy}{dt}$$

Substituting for dy/dt using equation [2] gives

$$\frac{d^2x}{dt^2} = -0.1\frac{dx}{dt} + 0.5(0.1x - 0.5y)$$

If we now substitute for y using equation [1] then

$$\frac{\mathrm{d}^2x}{\mathrm{d}t^2} = -0.1\frac{\mathrm{d}x}{\mathrm{d}t} + 0.05x - 0.25\left(2\frac{\mathrm{d}x}{\mathrm{d}t} + 0.2x\right)$$

$$\frac{\mathrm{d}^2x}{\mathrm{d}t^2} + 0.6\frac{\mathrm{d}x}{\mathrm{d}t} = 0$$

This second-order differential equation describes how x varies with time and can be solved. Thus with a solution of the form $x = A\,\mathrm{e}^{st}$ we have the auxiliary equation

$$s^2 + 0.6s = 0$$

$$s(s + 0.6) = 0$$

Thus the roots are $s = 0$ and -0.6 and so the solution is

$$x = A + B\mathrm{e}^{-0.6t}$$

The solution for y can be deduced in a similar way or by substituting the above value of x into one of the differential equations.

Example

Figure 9.2 shows the basic circuit of a transformer. Derive a differential equation showing how the current i_2 in the secondary circuit changes with time t when there is an input to the primary circuit of $V_m \cos \omega t$.

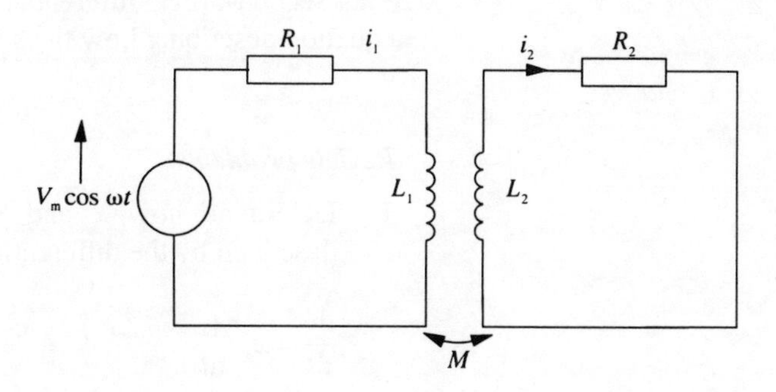

Fig. 9.2 Transformer

For the primary circuit we have, as a result of applying Kirchhoff's voltage law,

$$i_1R_1 + L_1\frac{\mathrm{d}i_1}{\mathrm{d}t} + M\frac{\mathrm{d}i_2}{\mathrm{d}t} = V_m \cos \omega t \qquad [3]$$

and for the secondary circuit

$$i_2 R_2 + L_2 \frac{di_2}{dt} + M \frac{di_1}{dt} = 0 \qquad [4]$$

Using equation [4] to substitute in equation [1] for di_1/dt gives

$$i_1 R_1 - \frac{L_1}{M}\left(L_2 \frac{di_2}{dt} + i_2 R_2\right) + M \frac{di_2}{dt} = V_m \cos \omega t$$

Differentiating with respect to t then gives

$$R_1 \frac{di_1}{dt} - \frac{L_1}{M}\left(L_2 \frac{d^2 i_2}{dt} + R_2 \frac{di_2}{dt}\right) + M \frac{d^2 i_2}{dt^2} = -\omega V_m \sin \omega t$$

Using equation [4] to substitute for di_1/dt gives

$$\frac{R_1}{M}\left(L_2 \frac{di_2}{dt} + i_2 R_2\right) - \frac{L_1}{M}\left(L_2 \frac{d^2 i_2}{dt} + R_2 \frac{di_2}{dt}\right)$$

$$+ M \frac{d^2 i_2}{dt^2} = -\omega V_m \sin \omega t$$

This can be rearranged to give

$$\left(M - \frac{L_1 L_2}{M}\right) \frac{d^2 i_2}{dt^2} + (R_1 L_2 - R_2 L_1) \frac{1}{M} \frac{di_2}{dt}$$

$$+ \frac{R_1 R_2}{M} i_2 = -\omega V_m \sin \omega t$$

This second-order differential equation can be solved to give an equation describing how the secondary current varies with time.

Review problems

1 Determine how x and y vary with time t for a system described by the differential equations:

$$\frac{dy}{dt} = x, \quad \frac{dx}{dt} = y$$

2 Determine how x and y vary with time t for a system described by the differential equations:

$$\frac{dy}{dt} + 3y - 4x = 0, \quad \frac{dx}{dt} + y + 8x = 0$$

3 Determine how x and y vary with time t for a system described by the differential equations:

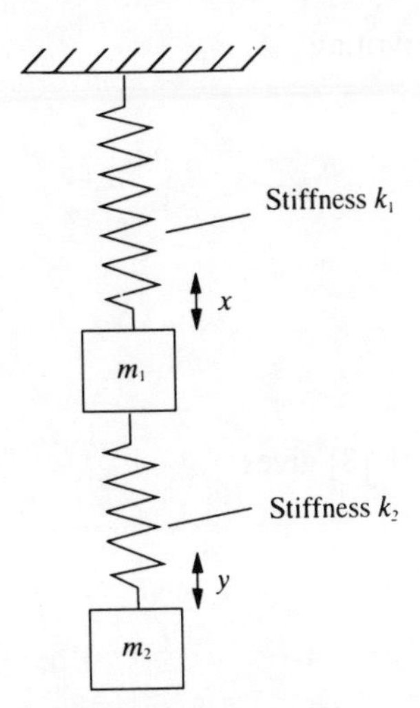

Fig. 9.3 Problem 5

$$\frac{dy}{dt} - 5y - 2x = 0, \ \frac{dx}{dt} + y - 7x = 0$$

4 Determine how x and y vary with time t for a system described by the differential equations:

$$\frac{dy}{dt} = -12x, \ \frac{dx}{dt} = 3y$$

if at $t = 0$ we have $x = 3$ and $y = 4$.

5 Determine a pair of simultaneous differential equations which can describe the motion of the two masses shown in figure 9.3.

9.2 Transforming higher order equations

Consider a simple situation of a freely falling object. The motion can be described as being a constant acceleration of 10 m/s^2, the approximate value for the acceleration due to gravity. Thus we can write for the differential equation describing how the displacement x varies with time t,

$$\frac{d^2x}{dt^2} = 10 \qquad [5]$$

We can transform this equation into two first-order differential equations. Thus, let $v = dx/dt$. Then $dv/dt = d^2x/dt^2$ and so the second-order differential equation can be replaced by two first-order differential equations, namely,

$$\frac{dv}{dt} = 10 \qquad [6]$$

and

$$v = \frac{dx}{dt} \qquad [7]$$

Thus we can solve the second-order differential equation by solving the two simultaneous first-order differential equations. The solution to equation [6] is

$$v = 10t + A \qquad [8]$$

To solve equation [7] we need to rearrange it so that it includes only two variables. This can be done by writing

$$v = \frac{dx}{dt} = \frac{dx}{dv}\frac{dv}{dt} = 10\frac{dx}{dv}$$

Thus

$$\frac{v^2}{2} = 10x + B$$

Hence eliminating v by the use of equation [8] gives

$$\tfrac{1}{2}(10t + A)^2 = 10x + B$$

$$x = 5t^2 + Ct + D$$

where C and D are constants.

Any second-order differential equation can be transformed into a pair of simultaneous first-order differential equations. A third-order differential equation can be transformed into three simultaneous first-order differential equations, a fourth-order differential equation into four simultaneous first-order differential equations. This form of transformation enables higher order differential equations to be solved by the numerical methods given in chapter 5 for first-order differential equations. It also enables non-linear differential equations to be tackled.

Example

Transform the following second-order, non-linear differential equation into a pair of simultaneous first-order differential equations.

$$\frac{d^2y}{dx^2} + 4y\frac{dy}{dx} = 0$$

Let $v = dy/dx$, then

$$\frac{d^2y}{dx^2} = \frac{dv}{dx} = \frac{dv}{dy}\frac{dy}{dx} = v\frac{dv}{dy}$$

The differential equation thus becomes

$$v\frac{dv}{dy} + 4yv = 0$$

and so the pair of simultaneous differential equations are

$$\frac{\mathrm{d}v}{\mathrm{d}y} + 4y = 0$$

and $v = \mathrm{d}y/\mathrm{d}x$.

Review problems

6 Transform the following second-order differential equation into a pair of first-order simultaneous differential equations.

$$\frac{\mathrm{d}^2 y}{\mathrm{d}x^2} = -y$$

7 Transform the following second-order differential equation into a pair of first-order simultaneous differential equations.

$$\frac{\mathrm{d}^2 y}{\mathrm{d}x^2} + 3\frac{\mathrm{d}y}{\mathrm{d}x} = 2$$

Further problems

8 Determine how x and y vary with time t for the system described by the following differential equations:

$$\frac{\mathrm{d}y}{\mathrm{d}t} = -x, \ \frac{\mathrm{d}x}{\mathrm{d}t} = y$$

9 Determine how x and y vary with time t for the system described by the following differential equations:

$$\frac{\mathrm{d}y}{\mathrm{d}t} + 5y + 2x = \mathrm{e}^t, \ \frac{\mathrm{d}x}{\mathrm{d}t} + 3y + 4x = t$$

10 Determine how x and y vary with time t for the system described by the following differential equations:

$$-3\frac{\mathrm{d}y}{\mathrm{d}x} + 3y + 4x = 3t, \ 3\frac{\mathrm{d}x}{\mathrm{d}t} + 2y + 3x = \mathrm{e}^t$$

11 Derive a differential equation showing how the current i varies with time for the circuit shown in figure 9.4. Hint: you

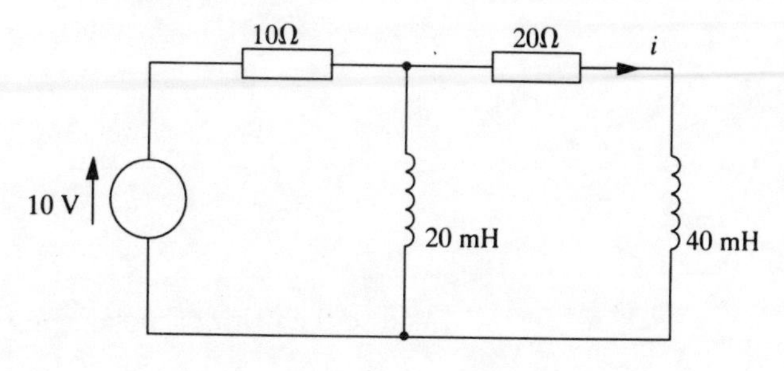

Fig. 9.4 Problem 11

could use mesh analysis to arrive at differential equations for each mesh and then solve the resulting simultaneous first-order differential equations.

12 A radioactive substance X decays to another radioactive substance Y which in turns decays into a stable substance Z. With x, y and z as the amounts of the substances at time t then the rates at which these amounts are changing are given by

$$\frac{dx}{dt} = -\lambda x, \ \frac{dy}{dt} = \lambda x - \mu y, \ \frac{dz}{dt} = \mu y$$

where λ and μ are constants. Derive equations showing how the amounts of X, Y and Z change with time if at $t = 0$ we have $x = A$, $y = 0$ and $z = 0$.

13 The currents i_1 and i_2 in two coupled circuits are given by the differential equations

$$L\frac{di_1}{dt} + Ri_1 + R(i_1 - i_2) = V, \ L\frac{di_2}{dt} + Ri_2 - R(i_1 - i_2) = 0$$

At time $t = 0$ we have both currents zero. Derive equations showing how the currents vary with time.

14 For the transformer circuit shown in figure 9.2 derive equations showing how the secondary current varies with time if $R_1 = 50\ \Omega$, $L_1 = 0.2$ H, $R_2 = 240\ \Omega$, $L_2 = 0.4$ H, $M = 0.2$ H and the input to the primary circuit is $170 \sin 900t$.

15 Transform the following differential equations into simultaneous first-order differential equations:

$$\text{(a)}\ \frac{d^2y}{dx^2} + 5\frac{dy}{dx} + 2 = 0, \ \text{(b)}\ x^2\frac{d^2y}{dx^2} + x\frac{dy}{dx} = 5$$

10 The differential operator

10.1 The D operator

When we write $\mathrm{d}y/\mathrm{d}x$ we think of the $\mathrm{d}/\mathrm{d}x$ as implying that the operation of differentiation with respect to x has to be carried out on y. Suppose in writing $\mathrm{d}y/\mathrm{d}x$ we replace $\mathrm{d}/\mathrm{d}x$ by D. Thus

$$\mathrm{D}y = \frac{\mathrm{d}}{\mathrm{d}x}(y) \qquad [1]$$

Since

$$\frac{\mathrm{d}^2 y}{\mathrm{d}x^2} = \frac{\mathrm{d}}{\mathrm{d}x}\left\{\frac{\mathrm{d}}{\mathrm{d}x}(y)\right\}$$

then we can write

$$\frac{\mathrm{d}^2 y}{\mathrm{d}x^2} = \mathrm{D}\{\mathrm{D}(\mathrm{y})\} = \mathrm{D}^2 y \qquad [2]$$

Similarly

$$\frac{\mathrm{d}^3 y}{\mathrm{d}x^3} = \mathrm{D}^3 y \qquad [3]$$

$$\frac{\mathrm{d}^4 y}{\mathrm{d}x^4} = \mathrm{D}^4 y \qquad [4]$$

and so on for yet higher orders.

When we have an expression of the form

$$\frac{\mathrm{d}y}{\mathrm{d}x} + ay$$

then using the differential operator notation we can write $\mathrm{D}y + ay$. However, if we treat this as a simple algebraic expression, this can be written as

$$\mathrm{D}y + ay = (\mathrm{D} + a)y \qquad [5]$$

The operator acting on y is now $(\mathrm{D} + a)$.

When we have an expression of the form

$$\frac{\mathrm{d}}{\mathrm{d}x}(ay)$$

then using the differential operator notation we can write $\mathrm{D}(ay)$. If a is a constant then since $\mathrm{d}(ay)/\mathrm{d}x = a\,\mathrm{d}y/\mathrm{d}x$ we have

$$\mathrm{D}(ay) = a\,\mathrm{D}y \qquad [6]$$

This expression is only true if a is a constant.

When we have an expression of the form

$$\frac{\mathrm{d}}{\mathrm{d}x}(u + v)$$

then using the differential operator notation we can write $\mathrm{D}(u + v)$. Since

$$\frac{\mathrm{d}}{\mathrm{d}x}(u + v) = \frac{\mathrm{d}u}{\mathrm{d}x} + \frac{\mathrm{d}v}{\mathrm{d}x}$$

then

$$\mathrm{D}(u + v) = \mathrm{D}u + \mathrm{D}v \qquad [7]$$

Example

Write the following differential equation in D notation:

$$\frac{\mathrm{d}^2y}{\mathrm{d}t^2} + 5\frac{\mathrm{d}y}{\mathrm{d}x} + 3y$$

Taking it term by term we have

$$\mathrm{D}^2y + 5\mathrm{D}y + 3y$$

This can then be written as

$$(\mathrm{D}^2 + 5\mathrm{D} + 3)y$$

Example

Write the following differential equation in D notation:

$$\frac{\mathrm{d}^2y}{\mathrm{d}x^2} + 5\frac{\mathrm{d}y}{\mathrm{d}x} + 6y = x^2$$

Taking it term by term we have

$$D^2y + 5Dy + 6y = x^2$$

$$(D^2 + 5D + 6)y = x^2$$

$$(D + 3)(D + 2)y = x^2$$

$$y = \frac{x^2}{(D+3)(D+2)}$$

Example

Determine the value of $D^2(5x^2 + 3)$.

This can be written as $D^2(5x^2) + D^2 3$. Since the differentiation of a constant, the 3, gives 0 then $D^2 3 = 0$. Since the differentiation of $5x^2$ is $10x$ and a second differentiation gives 10 then $D^2(5x^2) = 10$. Thus the value of the expression is 10.

Review problems

1 Write the following differential equations in D notation:

$$\text{(a) } \frac{dy}{dx} + 2y, \text{ (b) } \frac{d^2y}{dx^2} + 2\frac{dy}{dx} + 7y, \text{ (c) } \frac{d^3y}{dx^3} + 2\frac{d^2y}{dx^2} + 5\frac{dy}{dx} + 3y,$$

$$\text{(d) } \frac{d^2y}{dx^2} + 3\frac{dy}{dx} + 2y = 4x, \text{ (e) } \frac{d^2y}{dx^2} + 8\frac{dy}{dx} + 12y = 5x^2 + 3$$

2 Determine the values of the following:

(a) $D(x^2)$, (b) $D(2x^2 + 3)$, (c) $D^2(4x^2)$, (d) $D^2(2x^3+1)$

10.1.1 The inverse operator

Since we can write

$$\frac{d}{dx}\left(\int y\,dx\right) = y$$

then

$$D\left(\int y\,dx\right) = y$$

and so, applying the normal rules of algebra,

$$\int y\,dx = \frac{y}{D} = D^{-1}y \qquad [8]$$

In a similar way it can be shown that

$$\int\left(\int y\,\mathrm{d}x\right)\mathrm{d}x = \mathrm{D}^{-2}y \tag{9}$$

and so on for further powers.

Example

Determine the value of

$$\frac{1}{1-\mathrm{D}}x^2$$

This can be written as $(1-\mathrm{D})^{-1}x^2$ and expanded by means of the binomial theorem, or long division, to give

$$(1+\mathrm{D}+\mathrm{D}^2+\ldots)x^2 = x^2+\mathrm{D}x^2+\mathrm{D}^2x^2+\ldots$$

$$= x^2+2x+2$$

Review problems

3 Determine the values of the following:

$$\text{(a) } \frac{1}{1+\mathrm{D}}(1+x^3),\ \text{(b) } \frac{1}{(\mathrm{D}-1)(\mathrm{D}-2)}x^2$$

10.1.2 Exponentials

Since

$$\frac{\mathrm{d}}{\mathrm{d}x}(\mathrm{e}^{ax}) = a\,\mathrm{e}^{ax}$$

then

$$\mathrm{D}(\mathrm{e}^{ax}) = a\,\mathrm{e}^{ax} \tag{10}$$

Also, since

$$\frac{\mathrm{d}^2}{\mathrm{d}x^2}(\mathrm{e}^{ax}) = a^2\,\mathrm{e}^{ax}$$

then

$$\mathrm{D}^2(\mathrm{e}^{ax}) = a^2\,\mathrm{e}^{ax} \tag{11}$$

and so on for higher powers of D.

Suppose we have $(\mathrm{D}^2+2)\mathrm{e}^{3x}$, then we can write this as

$$D^2(e^{3x}) + 2\,e^{3x} = 3^2 e^{3x} + 2\,e^{3x}$$

This can be written as $(3^2 + 2)e^{3x}$. What we have effectively done is just replace the D by the a from the e^{ax}. In general, we can write this relationship as

$$f(D)e^{ax} = f(a)e^{ax} \qquad [12]$$

i.e. if we have some function of D operating on an exponential to the power ax then the result is the same function but of a multiplied by the exponential.

In a similar way we can derive relationships for the inverse operator. Thus, since

$$\int e^{ax}\,dx = \frac{1}{a}e^{ax}$$

then

$$\frac{1}{D}e^{ax} = \frac{1}{a}e^{ax} \qquad [13]$$

In general we can write

$$\frac{1}{f(D)}e^{ax} = \frac{1}{f(a)}e^{ax} \qquad [14]$$

There are some situations where using the above equation results in a value of infinity. Thus, for example,

$$\frac{1}{D-2}e^{2x}$$

gives

$$\frac{1}{2-2}e^{2x}$$

In such a case the problem has to be tackled in a different way. We can consider the e^{2x} to be $1 \times e^{2x}$ and use a method for the product of two terms. Thus, since

$$\frac{d}{dx}(uv) = u\frac{dv}{dx} + v\frac{du}{dx}$$

$$\frac{d}{dx}(1 \times e^{ax}) = 1 \times a\,e^{ax} + e^{ax}\frac{d}{dx}1$$

$$= e^{ax}(D+a)1$$

Hence

$$\frac{1}{D-2}e^{2x} = e^{2x}\frac{1}{(D+2)-2}1$$

$$= e^{2x}D^{-1}1 = xe^{2x}$$

In general, when we are concerned with the product of some variable V and an exponential,

$$D(Ve^{ax}) = e^{ax}(D + a)V \qquad [15]$$

and for higher orders

$$D^n(Ve^{ax}) = e^{ax}(D + a)V \qquad [16]$$

and the inverse operator

$$\frac{1}{f(D)}(Ve^{ax}) = e^{ax}\frac{1}{f(D+a)}V \qquad [17]$$

Example

Determine the value of

$$\frac{1}{D^2+5D+2}e^{-2x}$$

We have some function of D operating on an exponential to the power ax. The result is the same function but of a multiplied by the exponential (equation [14]), hence

$$\frac{1}{(-2)^2+5(-2)+2}e^{-2x} = -\frac{1}{4}e^{-2x}$$

Example

Determine the value of

$$\frac{1}{D(D-3)}e^{3x}$$

The usual method cannot be used here since it would lead to a value of infinity. Thus we rewrite the equation as

$$\frac{1}{D(D-3)}1 \times e^{3x}$$

Then, using equation [17], this becomes

$$e^{3x}\frac{1}{(D+3)(D+3-3)}1 = e^{3x}\frac{1}{(D+3)D}1$$

Using partial fractions this becomes

$$\frac{e^{3x}}{3}\left(\frac{1}{D} - \frac{1}{D+3}\right)1 = \frac{1}{3}e^{3x}\left[D^{-1} - \frac{1}{3}\left(1 - \frac{1}{3}D + \ldots\right)\right]1$$

$$= \frac{1}{3}e^{3x}\left(x - \frac{1}{3}\right)$$

Review problems

4 Determine the values of the following:

(a) $\frac{1}{D^2 + D + 1}e^{-2x}$, (b) $\frac{1}{D^2 + 3D + 3}e^{2x}$, (c) $\frac{1}{D^3}x^2 e^x$,

(d) $\frac{1}{D^2 - D - 2}e^{2x}$, (e) $\frac{1}{D^2 + 2D - 3}4e^{-x}$

10.1.3 Sines and cosines

Since we have

$$\frac{d}{dx}(\sin ax) = a\cos ax$$

$$\frac{d^2}{dx^2}(\sin ax) = -a^2 \sin ax$$

$$\frac{d^3}{dx^3}(\sin ax) = -a^3 \sin ax$$

$$\frac{d^4}{dx^4}(\sin ax) = a^4 \sin ax$$

Then we can write

$$D^2 \sin ax = (-a^2)\sin ax \qquad [18]$$

$$(D^2)^2 \sin ax = (-a^2)^2 \sin ax \qquad [19]$$

and so on. In general we can write

$$f(D^2)^n \sin(ax + b) = f(-a^2)^n \sin(ax + b) \qquad [20]$$

Likewise

$$\mathrm{D}^2 \cos ax = (-a^2)\cos ax \qquad [21]$$

$$(\mathrm{D}^2)^2 \cos ax = (-a^2)^2 \cos ax \qquad [22]$$

In general we can write

$$f(\mathrm{D}^2)^n \cos(ax+b) = f(-a^2)^n \cos(ax+b) \qquad [23]$$

We can develop similar relationships for the inverse operator. Thus, since

$$\int \sin ax \, \mathrm{d}x = -\frac{1}{a}\cos ax$$

$$\int\left(\int \sin ax \, \mathrm{d}x\right)\mathrm{d}x = -\frac{1}{a^2}\sin ax$$

Thus

$$\frac{1}{\mathrm{D}^2}\sin ax = -\frac{1}{a^2}\sin ax \qquad [24]$$

In general we can write

$$\frac{1}{f(\mathrm{D}^2)}\sin(ax+b) = -\frac{1}{f(-a^2)}\sin(ax+b) \qquad [25]$$

Similarly

$$\frac{1}{\mathrm{D}^2}\cos ax = -\frac{1}{a^2}\cos ax \qquad [26]$$

and in general

$$\frac{1}{f(\mathrm{D}^2)}\cos(ax+b) = -\frac{1}{f(-a^2)}\cos(ax+b) \qquad [27]$$

In some cases the above method of writing $-a^2$ for D^2 results in a zero in the denominator and so a value of infinity, e.g.

$$\frac{1}{\mathrm{D}^2+4}\cos 2x \rightarrow \frac{1}{-2^2+4}\cos 2x$$

Such situations can be handled in a number of ways. One way is to use the Euler relationship and transform the sine into an exponential.

Example

Determine the value of

$$\frac{1}{D^2 - 2}\sin 4x$$

Using equation [27], we have for the value

$$\frac{1}{(-4^2) - 2}\sin 4x = -\frac{1}{18}\sin 4x$$

Example

Determine the value of

$$\frac{1}{D^2 - D + 2}\sin 2x$$

We can use equation [27] to replace the D^2 by -2^2 and so give

$$\frac{1}{-4 - D + 2}\sin 2x$$

This can then be rewritten as

$$-\frac{1}{D + 2}\sin 2x$$

and multiplying the top and bottom by (D – 2) gives

$$-\frac{D - 2}{D^2 - 4}\sin 2x$$

A further replacement of D^2 by -2^2 gives

$\frac{1}{8}(D - 2)\sin 2x$

which can then be evaluated to give

$$\tfrac{1}{8}(2\cos 2x - 2\sin 2x) = \tfrac{1}{4}(\cos 2x - \sin 2x)$$

Review problems

5 Determine the values of the following:

$$\text{(a)}\ \frac{1}{D^2 + 1}\sin 3x,\ \text{(b)}\ \frac{1}{2D^2 - 3}\sin x,$$

$$\text{(c)}\ \frac{5}{D^3 - D^2 + 13D - 6}\sin 3x$$

10.2 Solving differential equations

The following illustrates how non-homogeneous differential equations can be solved by using the D operator to determine the particular integral.

Consider the solving of the non-homogeneous differential equation

$$\frac{d^2y}{dx^2} - \frac{dy}{dx} - 2y = e^x$$

The solution can be considered to be the sum of the complementary function and the particular integral. The complementary function is the solution of the corresponding homogeneous differential equation. Thus if we take the complementary solution to be of the form $y_c = A\,e^{sx}$ then the auxiliary equation is

$$s^2 - s - 2 = 0$$

$$(s-2)(s+1) = 0$$

Thus we have $s_1 = 2$ and $s_2 = -1$. Hence

$$y_c = A\,e^{2x} + B\,e^{-x}$$

The particular integral is a solution of the non-homogeneous differential equation. This can be written as

$$\left(D^2 - D - 2\right)y_p = e^x$$

and so

$$y_p = \frac{1}{D^2 - D - 2}\,e^x$$

Thus, using equation [15],

$$y_p = \frac{1}{1^2 - 1 - 2}\,e^x = -\frac{1}{2}\,e^x$$

Hence the solution is

$$y = y_c + y_p = A\,e^{2x} + B\,e^{-x} - \tfrac{1}{2}\,e^x$$

Review problems

6 Solve the following differential equations:

(a) $\frac{d^2y}{dx^2} - 2\frac{dy}{dx} - 3y = e^{2x}$, (b) $\frac{d^2y}{dx^2} - 3\frac{dy}{dx} - 4y = x^2 - 2x$,

(c) $\dfrac{d^2y}{dx^2} + 9y = 10\sin 2x$, (d) $\dfrac{d^2y}{dx^2} + \dfrac{dy}{dx} + y = e^x$,

(e) $\dfrac{d^2y}{dx^2} - 3\dfrac{dy}{dx} + 6y = x^2$, (f) $\dfrac{d^2y}{dx^2} + \dfrac{dy}{dx} = x^3 + 1$

10.3 Simultaneous differential equations

The following example illustrates how simultaneous differential equations can be tackled when they are written in terms of the D operator.

Suppose we have two simultaneous differential equations

$$3\frac{dx}{dt} + 3x + 2y = e^t, \; 4x - 3\frac{dy}{dt} + 3y = 3t$$

Writing these in D notation with D representing d/dt gives

$$3(D+1)x + 2y = e^t \qquad [28]$$

$$4x - 3(D-1)y = 3t \qquad [29]$$

We can eliminate x by using the value given by the second equation and substituting for it in the first equation. Thus

$$\tfrac{3}{4}(D+1)[3t + 3(D-1)y] + 2y = e^t$$

$$\tfrac{9}{4} + \tfrac{9}{4}t + \tfrac{9}{4}(D^2-1)y + 2y = e^t$$

$$(9D^2 - 1)y = 4e^t - 9t - 9$$

and so the particular integral is given by

$$y = \frac{1}{9D^2 - 1}(4e^t - 9t - 9)$$

$$= \tfrac{1}{2}e^t + 9t + 9$$

The homogeneous form of the differential equation is

$$\left(3D^2 - 1\right)y = 0$$

This gives an auxiliary equation of

$$(3s^2 - 1) = 0$$

and so the complementary function of $A\,e^{t/3} + B\,e^{-t/3}$. Thus the solution is

$$y = A\,e^{t/3} + B\,e^{-t/3} + \tfrac{1}{2}e^{t} + 9t + 9$$

The value of x can be obtained by substituting this value in equation [29]. Thus

$$4x - 3(D-1)(A\,e^{t/3} + B\,e^{-t/3} + \tfrac{1}{2}e^{t} + 9t + 9) = 3t$$

$$4x - 3(\tfrac{1}{3}A\,e^{t/3} - \tfrac{1}{3}B\,e^{-t/3} + \tfrac{1}{2}e^{t} + 9 - A\,e^{t/3} - B\,e^{-t/3} - \tfrac{1}{2}e^{t} - 9t - 9) = 3t$$

$$x = -\tfrac{1}{2}A\,e^{t/3} - B\,e^{-t/3} - 6t$$

Review problems

7 Solve the following simultaneous differential equations:

(a) $3\dfrac{dx}{dt} + 3x + 2y = e^{t}$, $4x - 3\dfrac{dy}{dt} + 3y = 0$;

(b) $\dfrac{dx}{dt} + \dfrac{dy}{dt} - 2x - 4y = e^{t}$, $\dfrac{dx}{dt} + \dfrac{dy}{dt} - y = e^{4t}$;

(c) $\dfrac{dx}{dt} + \dfrac{dy}{dt} - 3x - 15y = -4t$, $\dfrac{dx}{dt} + 2\dfrac{dy}{dt} + x = 5t^2$

8 Solve the following simultaneous differential equations when $x = 1$ and $y = 0$ at $t = 0$:

$$\frac{dx}{dt} + 2x + y = 0, \quad \frac{dy}{dt} + x + 2y = 0$$

Further problems

9 Write the following differential equations in D notation:

(a) $\dfrac{d^2y}{dx^2} - 3\dfrac{dy}{dx} - 4y = 7$, (b) $\dfrac{d^2y}{dx^2} + 6\dfrac{dy}{dx} + 5y = 3x^2$

10 Determine the values of the following:

(a) $D(x^2 + 2x + 3)$, (b) $\dfrac{1}{D(D+1)}(1 + x^3)$, (c) $\dfrac{1}{D^2 + 2D + 3}x^2$,

(d) $\dfrac{1}{D+3}e^{2x}$, (e) $\dfrac{1}{D^2+2D+5}e^{-x}$, (f) $\dfrac{1}{D^2+9}10\sin 2x$,

(g) $\dfrac{1}{D^2-3D-4}5\sin 2x$

11 Solve the following differential equations:

(a) $\dfrac{d^2y}{dx^2}-3\dfrac{dy}{dx}+2y=8x-16$, (b) $\dfrac{d^2y}{dx^2}-\dfrac{dy}{dx}-6y=e^x$,

(c) $\dfrac{d^2y}{dx^2}+5\dfrac{dy}{dx}+6y=3x+5$, (d) $\dfrac{d^2y}{dx^2}-4\dfrac{dy}{dx}+4y=4\cos 2x$

12 Solve the following simultaneous differential equations:

(a) $\dfrac{dx}{dt}+2y+5x=e^t$, $\dfrac{dy}{dt}+4y+3x=t$,

(b) $\dfrac{dx}{dt}+4x+6y=3e^t$, $\dfrac{dy}{dt}+5y+2x=4$,

(c) $\dfrac{dx}{dt}-\dfrac{dy}{dt}-2x+4y=t$, $\dfrac{dx}{dt}+\dfrac{dy}{dt}-x-y=1$,

(d) $\dfrac{dx}{dt}-x-2y=0$, $\dfrac{dy}{dt}-3x+4y=0$

13 An electrical circuit consists of an inductance L in parallel with a resistance R. When a current source of $I\sin\omega t$ is applied to the circuit there is a current x through the inductance and a current y through the resistance. At time t = 0 both x and y are 0. Derive equations showing how the two currents vary with time.

14 An electrical circuit consists of an inductance L in parallel with a capacitance C. When a current source of $I\sin\omega t$ is applied to the circuit there is a current x through the inductance and a current y through the capacitance. At time t = 0 both x and y are zero. Derive equations showing how the two currents vary with time.

15 A particle moving in a plane has accelerations in the x and y directions given by

$$\frac{d^2x}{dt^2}=\omega\frac{dy}{dt} \text{ and } \frac{d^2y}{dt^2}=a-\omega\frac{dx}{dt}$$

When $t=0$ both x and y are 0. Determine equations showing how the displacements x and y of the particle vary with time t.

Answers

Chapter 1

1 (a) $v = 4t + 3$, (b) $v = 6\cos 2t$, (c) $v = -10\,e^{-2t}$

2 $\dfrac{d\theta}{dI} = 2cI$

3 (a) $\dfrac{dx}{dt} = 10t + 2$, $\dfrac{d^2x}{dt^2} = 10$, (b) $\dfrac{dx}{dt} = 6t^2 + 6t + 4$, $\dfrac{d^2x}{dt^2} = 12t + 6$, (c) $\dfrac{dx}{dt} = 12\cos 3t$, $\dfrac{d^2x}{dt^2} = -36\sin 3t$, (d) $\dfrac{dx}{dt} = -6\,e^{-3t}$, $\dfrac{d^2x}{dt^2} = 18\,e^{-3t}$

4 Negative

5 (a) $v = L\dfrac{di}{dt}$, (b) $F = c\dfrac{dx}{dt}$, (c) $i = \dfrac{dq}{dt}$, (d) $T = I\dfrac{d\omega}{dt}$, (e) $\dfrac{dV}{dp} = -\dfrac{V}{p}$, (f) $\dfrac{dx}{dt} = k\sqrt{\theta}$

6 (a) $m\dfrac{dv}{dt} + kv^2 = mg$, (b) $m\dfrac{dv}{dt} + kv = mg$, (c) $L\dfrac{di}{dt} + Ri = V$, (d) $A\dfrac{dh}{dt} + \sqrt{2gh} = 0$, (e) $\dfrac{dV}{dt} - (4\pi)^{1/3}3^{2/3}kV^{2/3} = 0$, (f) $m\dfrac{d^2x}{dt^2} + c\dfrac{dx}{dt} + kx = 0$

7 As given in the problem

8 (a) $A = 2$, (b) $A = 3/\omega$, $B = 4$, (c) $A = 2$, (d) $A = 5$

9 (a) $\dfrac{d^2y}{dx^2} + 2y = 0$, (b) $3\dfrac{d^2y}{dx^2} + 2\dfrac{dy}{dx} + 4y = 0$

10 (a) Order 1, degree 1, (b) order 2, degree 1, (c) order 2, degree 2, (d) order 2, degree 1

11 (a) Linear, (b) non-linear, (c) non-linear, (d) linear

12 (a) $\dfrac{dx}{dt} = 10t + 3$, (b) $\dfrac{d^2x}{dt^2} = 10$

13 $\dfrac{dq}{dt} = -20\,e^{-2t}$

14 $\dfrac{di}{dt} = 10\cos 5t$

15 (a) $i = C\dfrac{dv}{dt}$, (b) $q_1 - q_2 = C\dfrac{dp}{dt}$, (c) $\dfrac{dN}{dt} = kN$,
(d) $\dfrac{dN}{dt} = -kN$, (e) $\dfrac{dI}{dx} = -kx$

16 (a) $EI\dfrac{d^2y}{dx^2} = wL - wx$, (b) $m\dfrac{dv}{dt} = mg - \frac{4}{3}\pi r^3 \rho g - 6\pi\eta r v$,
(c) $v_C + RC\dfrac{dv_C}{dt} = 0$, (d) $m\dfrac{dv}{dt} + kv = 0$

17 (a) $y = x^2 + 4$, (b) $y = 3x^2 + 2x + 5$, (c) $y = 2 - \cos 3t$,
(d) $y = 4\,e^{-3t} + 6\,e^{4t}$

18 $A = 0$, $B = 0$

19 $A = -V$

20 (a) Order 1, degree 1, linear, (b) order 3, degree 1, linear, (c) order 2, degree 1, non-linear, (d) order 1, degree 1, non-linear, (e) order 1, degree 1, non-linear, (f) order 2, degree 1, linear

21 (a), (b)

Chapter 2

1 (a) $y = x^5 + A$, (b) $y = \frac{1}{2}x^2 + A$, (c) $y = -\cos x + A$,
(d) $y = 5\ln x + A$, (e) $x = \ln(3 + y) + A$, (f) $y = -\dfrac{1}{x} + \ln x + A$,
(g) $y = 2\,e^{2x} + A$, (h) $y^2 + 2y = \frac{1}{3}x^3 + A$, (i) $-\dfrac{x}{y} = \ln x + A$

2 (a) $y = 3x - \frac{1}{2}x^2 + 1$, (b) $x = \ln(1 + \frac{1}{2}y)$, (c) $y = 2.72\,e^{-1/x}$,
(d) $y^2 + y = x^3$

3 $\theta = \theta_0\,e^{-kt}$

4 $T_1 = T_2\,e^{\mu\theta}$

5 $q = Q\,e^{-t/RC}$

6 $\dfrac{dM}{dt} = 0.1M$, $M = M_0\,e^{0.1t}$

7 (a) $\dfrac{dv}{dt} = -kv$, $v = C\,e^{-kt}$, (b) $v = 200\,e^{-1.8t}$ m/s

8 116 min

9 0.04%

10 (a) $y = 2 + A\,e^{-3x}$, (b) $y = \frac{1}{2} + \frac{1}{2}A\,e^{x^2}$, (c) $y = A\,e^{3x} - e^{2x}$,
(d) $y = 9(3x - 1 + 10\,e^{3x})$, (e) $y = 2 - 2\,e^{x-1}$

11 (a) $y = Ce^x + 2e^{3x}$, (b) $y = Ce^x + 4$, (c) $y = Ce^x - 6 - 4x$, (d) $y = Ce^x - \frac{4}{3}\sin 2x - \frac{8}{3}\cos 2x$

12 (a) $y^2 + 5y = \frac{1}{3}x^3 + A$, (b) $y = e^x + A$, (c) $y = \ln x - \frac{1}{2}x^2 + A$, (d) $y = -\frac{1}{2}\cos 2x + A$, (e) $\ln y = -x^2 + A$, (f) $y = x^2 - \cos x + A$, (g) $x = -\frac{1}{2}\ln(2 - y^2) + A$, (h) $\ln y = \sin x + A$, (i) $\ln\left(\frac{y}{x}\right) = -\frac{3}{4}\ln\left[1 + 2\left(\frac{y}{x}\right)^2\right] + A$, (j) $x = A(x - y)^2$

13 (a) $\ln(2y - 1) = 2(x - 1)$, (b) $y = 2e^x$, (c) $e^y = 2x^3 + e^2$, (d) $y = 2x^5$, (e) $x = \sin 2y + 2$, (f) $y = \frac{1}{4}\cos 2\theta + 2$, (g) $e^{2y} = 2e^x - 1$, (h) $\ln y = x^3$, (i) $y = -\left(\frac{x}{\ln x - 1 - \ln 1}\right)$

14 $R = R_0 e^{\alpha\theta}$

15 144.6 s

16 0.17

17 (a) $y = -2x - 1 + Ae^x$, (b) $y = \frac{2}{2.5}x^2 + Ax^{-1/2}$, (c) $y = 1 + A(1 + x^2)^{-3/2}$, (d) $y = -2x - 1 + Ae^{2x}$, (e) $y = \frac{x}{x-1}\ln x - \frac{1}{x-1} + \frac{Ax}{x-1}$, (f) $y = Ae^{-3x} + \frac{4}{3}$, (g) $y = x^4(4\ln x + A)$, (h) $y = \frac{1}{3}e^x + Ae^{-2x}$, (i) $y = Ae^{-x^2/2} - 2$

18 $i = \frac{t}{R} + \frac{L}{R^2}(e^{-Rt/L} - 1)$

19 $v = V(1 - e^{-t/RC})$

20 $x = a(1 - e^{-kt})$

21 $x = \frac{ab[1 - e^{-(a-b)kt}]}{a - b e^{-(a-b)kt}}$

22 (a) $y = Ce^{-2x} + x - \frac{1}{2}$, (b) $y = Ce^{-2x} + \frac{1}{3}e^x$, (c) $y = Ce^{-2x} + x - 4$, (d) $y = Ce^{-2x} + \frac{1}{4} - \frac{1}{2}x + \frac{1}{2}x^2$

Chapter 3

1 $4\frac{dv_C}{dt} + v_C = 5$

2 $4\frac{dv_C}{dt} + v_C = 0$

3 (a) 8 V, (b) 12 μA, (c) 12 V

4 (a) 10.9 V, (b) −0.54 mA

5 $v_C = 6e^{-2t/3}$ V

6 $v_C = 24\,e^{-t/2} + 2t - 24$ V

7 $v_C = 30\,e^{-t/2} + 2t - 24$ V

8 $0.1\frac{di}{dt} + 10i = 4$

9 $\frac{0.1}{50}\frac{di}{dt} + i = 0$

10 $i = 0.1(1 - e^{-400t})$ A

11 $i = 1(1 - e^{-3t})$ A

12 $i = \frac{t^2}{R} - \frac{2Lt}{R^2} + \frac{2L^2}{R^3} - \frac{2L^2}{R^3}e^{-Rt/L}$

13 (a) 50 ms, (b)(i) 2.53 A, (ii) 3.46 A

14 20 A/s

15 (a) $0.2\frac{dv_C}{dt} + v_C = 3$, (b) $2\frac{dv_C}{dt} + v_C = 0$, (c) $0.2\frac{di}{dt} + i = 0.4$

16 $v_C = 4\,e^{-t/8}$ V

17 $v_C = 4 + 8(1 - e^{-250t})$ V

18 $i = 1.5\,e^{-40t}$ A

19 $i = \frac{1}{R-L}e^{-t}(1 - e^{-Rt/L})$

20 80 A/s

21 0.693

22 (a) 2 s, (b) 6.3 V

Chapter 4

1 (a) $v = gt + u$, (b) $x = \frac{1}{2}gt^2 + ut$

2 $v = v_0\,e^{kt}$

3 $v = -\frac{v_0}{v_0 kt - 1}$

4 $v^2 = \left(u^2 + \frac{g}{k}\right)e^{-2kx} - \frac{g}{k}$, $\frac{1}{2k}\ln\left(1 + \frac{ku^2}{g}\right)$

5 7.8 s

6 $100\frac{dv}{dt} = 500 - 10v$, $v = 50(1 - e^{-0.1t})$, 50 m/s

7 $v = 8(1 - e^{-t/400})$

8 $\frac{v-40}{v+40} = -e^{-0.49t}$

9 As given in the problem

10 $v = 2.45(1 - e^{-8t})$

11 $v = \frac{4u}{(kt\sqrt{u} + 2)^2}$

Chapter 5

1 (a) x, y; 1, 1; 2, 2; 3, 3; 3, 4
(b) x, y; 1, 2; 1.1, 2.150; 1.2, 2.287; 1.3, 2.415; 1.4, 2.533
(c) x, y; 1, 3; 1.1, 3.400; 1.2, 3.863; 1.3, 4.395; 1.4, 5.002
(d) x, y; 0, 1; 0.2, 1.200, 0.4, 1.520; 0.6, 1.984; 0.8, 2.621

2 t(s), v(m/s); 0, 2; 0.1, 2.98; 0.2, 3.96; 0.3, 4.94; 0.4, 5.92

3 t, v_C; 0, 2; 0.1, 1.8; 0.2, 1.62; 0.3, 4.94; 0.4, 5.92

4 x, Euler, improved Euler; 0, 1, 1; 0.1, 1.000, 1.005; 0.2, 1.010, 1.020; 0.3, 1.030, 1.045; 0.4, 1.060, 1.080

5 x, Euler, improved Euler; 0, 1, 1; 0.2, 1.200, 1.260; 0.4, 1.520, 1.665; 0.6, 1.984, 2.248; 0.8, 2.621, 3.046

6 (a) 0.84, (b) 0.8375

7 x, y; 0, 1; 0.2, 1.2642; 0.4, 1.6755; 0.6, 2.2663; 0.8, 3.0765

8 2.0670

9 3.1487

10 (a) x, y; 0, 0; 0.1, 0; 0.2, 0.01; 0.3, 0.031; 0.4, 0.064
(b) x, y; 0, 0; 0.2, 0; 0.4, 0.040; 0.6, 0.128; 0.8, 0.274
(c) x, y; 0, 0; 0.1, 0.1; 0.2, 0.202; 0.3, 0.310; 0.4, 0.429
(d) x, y; 0, 1; 0.1, 1.2; 0.2, 1.39; 0.3, 1.57; 0.4, 1.74

11 t(s), i(A); 0, 0; 0.2, 1; 0.4, 1.90; 0.6, 2.71; 0.8, 3.44

12 x, Euler, improved Euler; 0, 1.00, 1.00; 0.1, 1.00, 1.01; 0.2, 1.02, 1.04; 0.3, 1.06, 1.09; 0.4, 1.12, 1.16

13 2.2949

14 0.4317

15 (a) 1.125, (b) 1.12722

16 (a) 0.915, (b) 0.91365

17 1.22297

18 0.04179

19 0.41927

20 0.50818

21 x, y; 0, 0; 0.2, 0.02140; 0.4, 0.09182; 0.6, 0.22211; 0.8, 0.42552; 1.0, 0.71825

22 (a) 9.52 m/s, (b) 32.97 m/s

23 4.65 m/s

24 (a) 35.73 m/s, (b) 22.81 m/s

25 1.967 V

Chapter 6

1 (a) $y = x^2 + 3x + 2$, (b) $y = 3x^2 + 2$, (c) $y = \frac{2}{3}x^3 + 2x + 1$, (d) $y = \frac{1}{4}x^4 + x^2$

2 $y = \dfrac{Fx^3}{6EI}(3L - x)$

3 (a) $y = A\,e^x + B\,e^{-x}$, (b) $y = A\,e^{2x} + B\,e^x$, (c) $y = A\,e^x + B\,e^{-5x}$

4 (a) $y = e^{-x}(Ax + B)$, (b) $y = e^{-3x}(Ax + B)$, (c) $y = e^x(A\cos 3x + B\sin 3x)$, (d) $y = e^{2x}(A\cos x + B\sin x)$, (e) $y = e^x(A\cos x + B\sin x)$, (f) $y = A\,e^{12.05x} + B\,e^{-3.05x}$, (g) $y = e^{-x}(A\cos 1.73x + B\sin 1.73x)$, (h) $y = e^{-4x}(Ax + B)$

5 (a) $y = 3x\,e^x$, (b) $y = (3x + 2)e^{-3x}$, (c) $y = e^{-2x}(\cos x - \sin x)$

6 (a) $y = A\,e^{-3x} + B\,e^{-x} + 3$, (b) $y = A\cos 2x + B\sin 2x + 2$, (c) $y = A + B\,e^{-4x} + 4x$, (d) $y = A\,e^{2x} + B\,e^{-2x} + 2x\,e^{2x}$, (e) $y = A\cos 2x + B\sin 2x + 6x^3 - 9x$, (f) $y = A\,e^x + B\,e^{3x} + 4x + 4$, (g) $y = A\,e^{-x} + B\,e^{2x} - 3\cos x - \sin x$, (h) $y = A\,e^{3x} + B\,e^{-x} - e^x + 2\sin x - \cos x$, (i) $y = A\,e^x + B\,e^{2x} + e^{-x} + \frac{7}{10}\cos 3x + \frac{9}{10}\sin 3x$

7 (a) $y = A + e^{-x}(B + Cx)$, (b) $y = (A + Bx) + (C\sin 2x + D\cos 2x)$, (c) $y = A + B\,e^{3x} + C\,e^{-2x} - \frac{3}{54}x^3 + \frac{15}{54}x^2 - \frac{10}{54}x$, (d) $y = A\,e^{-x} + Bx\,e^{-x} + Cx^2\,e^{-x} + x + 3$

8 (a) $y = x^2 + x$, (b) $y = \frac{2}{3}x^3 + 2x + 1$, (c) $y = \frac{1}{3}x^3 + \frac{1}{2}x^2$

9 $y = \dfrac{wx}{24EI}(x^3 - 2Lx^2 + L^3)$

10 $y = \dfrac{wx^2}{24EI}(6L^2 - 4Lx + x^2)$

11 (a) $y = A\,e^{2x} + B\,e^{-2x}$, (b) $y = A + B\,e^x$, (c) $y = A\,e^{5x} + B\,e^x$, (d) $y = e^{-2x}(Ax + B)$, (e) $y = A\,e^x + B\,e^{-5x}$, (f) $y = e^{-3x}(Ax + B)$, (g) $y = e^{3x}(A\cos 4x + B\sin 4x)$, (h) $y = e^{-3x}(A\cos x + B\sin x)$

12 (a) $y = \frac{1}{3}\sin 3x$, (b) $y = e^{2x}(3 - 5x)$, (c) $y = e^x\left(2\cos 3x - \frac{1}{3}\sin 3x\right)$

13 (a) $y = A\,e^x + B\,e^{2x} + \frac{1}{6}e^{-2x}$, (b) $y = A\,e^x + B\,e^{2x} + 2x\,e^{2x}$, (c) $y = A\,e^x + B\,e^{2x} + 2 + 6x + 8x^2$, (d) $y = A\,e^x + B\,e^{2x} - \sin 2x + 3\cos 2x$, (e) $y = A\,e^{-3x} + B\,e^{2x} + \cos 2x - \sin 2x$, (f) $y = A\,e^{5x} + B\,e^{-2x} + \frac{1}{7}x\,e^{-2x} - \frac{2}{5}$,

(g) $y = A\,e^{3x} + B\,e^{2x} - \frac{3}{10}\sin x - \frac{3}{10}\cos x$,
(h) $y = e^{x}(A\cos x + B\sin x) - \frac{1}{2}e^{x}\,x\cos x$

14 $y = \dfrac{k}{\omega_0^2 - \omega^2}(\cos\omega t - \cos\omega_0 t)$

15 $\dfrac{d^2x}{dt^2} + 4\dfrac{dx}{dt} + 3 = 0$, $x = 5\,e^{-t} - 2\,e^{-3t}$

16 $\dfrac{d^2x}{dt^2} = 0.2 + 0.2x$, $x = e^{\sqrt{0.2}\,t} + e^{-\sqrt{0.2}\,t} + 1$

17 $v = \frac{1}{200}V e^{-3t}(2\cos 4t - \sin 4t) + \frac{1}{100}V(4 - 5\,e^{-t})$

18 (a) $y = A\,e^{4x} + B\,e^{-x} + Cx\,e^{-x}$, (b) $y = A\,e^{3x} + Bx\,e^{3x} + Cx^2\,e^{3x}$,
(c) $y = A\,e^{2x} + Bx\,e^{2x} + Cx^2\,e^{2x}$,
(d) $y = A\,e^{x} + Bx\,e^{x} + Cx^2\,e^{x} - \frac{1}{8}e^{-x}$,
(e) $y = A\,e^{-x} + B\,e^{2x} + Cx\,e^{2x} + 2\,e^{x} - 2x\,e^{-x}$

Chapter 7

1 (a) 4 rad/s, (b) 4.32 m

2 (a) $y = \cos t$, (b) $y = \sqrt{2}\cos(t - 45°)$, (c) $y = \sqrt{2}\cos(t + 45°)$

3 $T = 2\pi\sqrt{\dfrac{m(k_1 + k_2)}{k_1 k_2}}$

4 $T = 2\pi\sqrt{\dfrac{m}{3k}}$

5 (a) $x = A\,e^{-4t} + B\,e^{-t}$, (b) $x = e^{-t}\left(C\cos\sqrt{3}\,t + D\sin\sqrt{3}\,t\right)$

6 6 N per m/s

7 (a) $x = \cos t$, (b) $x = e^{-t}(0.2t + 0.2)$,
(c) $x = e^{-2t}\left(\cos\sqrt{3}\,t + 0.67\sin\sqrt{3}\,t\right)$

8 $x = 4\,e^{-2t} - 2\,e^{-4t}$

9 1.38 Hz

10 $x = \cos 5t + \sin t$, 5 rad/s

11 $x = 2.5 \times 10^{-3}(\cos 10t + \cos 30t)$

12 $x = \sin t - 3\cos t$

13 $x = \cos 2t + 2\sin 2t$

14 $0.5\dfrac{d^2x}{dt^2} + 25x = 0$, 0.054 m

15 1.05 s, 0.091 m

16 $y = 2\pi\sqrt{\dfrac{mL^3}{48EI}}$

17 $y = 2\pi\sqrt{\dfrac{IL}{GJ}}$

18 (a) $x = e^{-t}(0.2\cos 2.24t + 0.089\sin 2.24t)$,
(b) $x = e^{-t}(0.2\cos 2.24t + 0.22\sin 2.24t)$
19 (a) No, (b) No
20 1000 N/m
21 $\theta = e^{-\alpha t}(A\cos\beta t + B\sin\beta t)$
22 $x = e^{-3t}(0.05\cos 12.56t + 0.12\sin 12.56t)$
23 $x = A\cos t + B\sin t - \cos 2t$
24 $x = \cos 3t - \cos 5t$
25 $x = A\cos 3t + B\sin 3t - \frac{15}{7}\cos 4t$
26 $x = e^{-t}(2\cos t - 6\sin t) - 2\cos 2t + 4\sin 2t$
27 2 rad/s, 3 rad/s
28 $x = e^{-0.25t}(A\cos 1.39t + B\sin 1.39t) + 4\cos t + 2\sin t$
29 $x = e^{-3t}(A\cos t + B\sin t) + \cos 2t + 2\sin 2t$
30 $x = e^{-t}(A\sin 3t + B\cos 3t) - \frac{12}{5}(3\sin 4t + 4\cos 4t)$

Chapter 8

1 $v_C = e^{-t}(10t + 10)$ V
2 Underdamped, 10 Ω
3 $v_C = 1 - e^{-4000t}(\cos 4899t + 4899\sin 4899t)$ V
4 Overdamped $(1/CR)^2 > 4/CL$, critically damped $(1/CR)^2 = 4/CL$, underdamped $(1/CR)^2 < 4/CL$
5 $v = 200\,e^{-200t}\sin 2449t$ V
6 62.5 Ω
7 $i_C = e^{-4000t}(4\sin 2000t - 2\cos 2000t)$ A
8 $i = e^{-4t}(10\cos 2t + 20\sin 2t)$ A
9 $i = 1 - t\,e^{-t} - e^{-t}$ A
10 $v_C = 20\,e^{-t} - 10\,e^{-2t}$ V
11 Overdamped
12 $(1/RC)^2 = 1/LC$
13 $v = 0.4 - e^{-3t}(0.4\cos 4t + 0.3\sin 4t)$ V
14 $\left(\frac{R_1}{2L} + \frac{1}{2R_2C}\right)^2 = \frac{R_1 + R_2}{R_2LC}$
15 $v = 24 - e^{-3t}(24\cos t + 32\sin t)$ V
16 $I = 5\,e^{-2t}\sin t$ A

Chapter 9

1 $x = A\,e^{t} + B\,e^{-t}$, $y = A\,e^{t} - B\,e^{-t}$

2 $y = A\,e^{-4t} + B\,e^{-7t}$, $x = C\,e^{-4t} + D\,e^{-7t}$

3 $y = e^{6t}(A\cos t + B\sin t)$, $x = e^{6t}(C\cos t + D\sin t)$

4 $y = 4\cos 6t - 6\sin 6t$, $x = 3\cos 6t + 2\sin 6t$

5 $m_1\dfrac{d^2x}{dt^2} = k_2(y-x) - k_1x$, $m_2\dfrac{d^2x}{dt^2} = -k_2(y-x)$

6 $v = \dfrac{dy}{dx}$, $\dfrac{dv}{dy} = -y$

7 $v = \dfrac{dy}{dx}$, $\dfrac{dv}{dx} + 3v = 2$

8 $x = A\cos t + B\sin t$, $y = B\cos t - A\sin t$

9 $x = A\,e^{-7t} + B\,e^{-2t} + \frac{5}{14} - \frac{31}{196} - \frac{1}{8}e^{t}$,

$y = A\,e^{-7t} + \frac{2}{3}B\,e^{-2t} + \frac{9}{98} - \frac{1}{7}t + \frac{5}{24}e^{t}$

10 $x = A\,e^{t/3} + B\,e^{-t/3} - 6t$, $y = -2A\,e^{t/3} - B\,e^{-t/3} + \frac{1}{2}e^{t} + 9t + 9$

11 $0.04\dfrac{d^2i}{dt^2} + 50\dfrac{di}{dt} + 10\,000i = 10$

12 $x = A\,e^{-\lambda t}$, $y = \dfrac{A\lambda}{\mu - \lambda}(e^{-\lambda t} - e^{-\mu t})$,

$z = \dfrac{A}{\mu - \lambda}(-\mu\,e^{-\lambda t} + \lambda\,e^{-\mu t} + \mu - \lambda)$

13 $i_1 = \dfrac{V}{R}\left(\frac{2}{3} - \frac{1}{2}e^{-Rt/L} - \frac{1}{6}e^{-3Rt/L}\right)$,

$i_2 = \dfrac{V}{R}\left(\frac{1}{3} - \frac{1}{2}e^{-Rt/L} + \frac{1}{6}e^{-3Rt/L}\right)$

14 $i_2 = 0.15\cos 900t - 0.45\sin 900t + \frac{1}{4}A\,e^{-200t} - \frac{5}{6}B\,e^{-1500t}$

15 (a) $v = \dfrac{dy}{dx}$, $\dfrac{dv}{dx} + 5v + 2 = 0$, (b) $v = \dfrac{dy}{dx}$, $x^2\dfrac{dv}{dx} + xv = 5$

Chapter 10

1 (a) $(D + 2)y$, (b) $(D^2 + 2D + 7)y$, (c) $(D^3 + 2D^2 + 5D + 3)y$,

(d) $y = \dfrac{4x}{(D+2)(D+1)}$, (e) $y = \dfrac{5x^2 + 3}{(D+6)(D+2)}$

2 (a) $2x$, (b) $4x$, (c) 8, (d) $12x$

3 (a) $x^3 - 3x^2 + 6x - 5$, (b) $\frac{1}{2}x^2 + \frac{3}{2}x + \frac{7}{4}$

4 (a) $\frac{1}{3}e^{-2x}$, (b) $\frac{1}{13}e^{2x}$, (c) $e^{x}(x^2 - 6x + 12)$, (d) $\frac{1}{3}x\,e^{2x}$, (e) $-e^{-x}$

5 (a) $-\frac{1}{8}\sin 3x$, (b) $-\frac{1}{5}\sin x$, (c) $\frac{5}{153}(3\sin 3x - 4\cos 3x)$

6 (a) $y = A\,e^{3x} + B\,e^{-x} - \frac{1}{3}e^{2x}$,

(b) $y = A\,e^{4x} + B\,e^{-x} - \frac{1}{4}x^2 + \frac{7}{8}x - \frac{25}{32}$,

(c) $y = A\cos 3x + B\sin 3x + 2\sin 2x$,

(d) $y = e^{-x/2}(A\cos 0.87x + B\sin 0.87x) + \frac{1}{3}e^x$,
(e) $y = Ae^x + Be^{2x} + \frac{1}{2}x^2 + \frac{3}{2}x + \frac{7}{4}$,
(f) $y = A + Be^{-x} + \frac{1}{4}x^4 - x^3 + 3x^2 - 5x + 5$

7 (a) $x = Ae^{t/3} + Be^{-t/3},\ y = -2Ae^{t/3} - Be^{-t/3} + \frac{1}{2}e^t$,
(b) $x = Ae^{-2t},\ y = -\frac{2}{3}Ae^{-2t} + \frac{1}{3}e^{4t} - \frac{1}{3}e^t$,
(c) $x = Ae^{-3t} + Be^{-5t} + 2 - 6t + 5t^2$,
$y = -\frac{1}{3}Ae^{-3t} - \frac{2}{5}Be^{-5t} - \frac{2}{3} + 2t - t^2$

8 $x = \frac{1}{2}(e^t + e^{-3t}),\ y = -\frac{1}{2}(e^{-t} - e^{-3t})$

9 (a) $y = \dfrac{7}{(D-4)(D+1)}$, (b) $y = \dfrac{3x^2}{(D+5)(D+1)}$

10 (a) $2x$, (b) $\frac{1}{4}x^4 - x^3 + 3x^2 - 5x + 5$, (c) $\frac{1}{3}x^2 - \frac{4}{9}x + \frac{2}{27}$, (d) $\frac{1}{5}e^{2x}$,
(e) $\frac{1}{4}e^{-x}$, (f) $2\sin 2x$, (g) $\frac{1}{12}(2\cos 2x - 8\sin 2x)$

11 (a) $y = Ae^x + Be^{2x} + 4x - 2$, (b) $y = Ae^{3x} + Be^{-2x} - \frac{1}{6}e^x$,
(c) $y = Ae^{-3x} + Be^{-2x} + 3x + \frac{9}{2}$, (d) $y = (Ax + B)e^{2x} + \frac{1}{2}\sin 2x$

12 (a) $y = Ae^{-7t} + Be^{-2t} + \frac{5}{14}t - \frac{31}{196} - \frac{1}{8}e^t$,
$x = Ae^{-7t} - \frac{2}{3}Be^{-2t} + \frac{5}{24}e^t - \frac{1}{7}t + \frac{9}{98}$,
(b) $x = Ae^{-t} + Be^{-8t} + e^{-t} - 3,\ y = -\frac{1}{2}Ae^{-t} + \frac{2}{3}Be^{-8t} - \frac{1}{3}e^t + 2$
(c) $x = Ae^{3t} + Be^t - \frac{1}{6}t - \frac{13}{18},\ y = Ae^{3t} + \frac{1}{3}Be^t + \frac{1}{6}t - \frac{15}{18}$,
(d) $x = Ae^{-5t} + Be^{2t},\ y = -3Ae^{-5t} + \frac{1}{2}Be^{2t}$

13 $x = \dfrac{IR}{\sqrt{R^2 + \omega^2 L^2}}\sin(\omega t - \phi) + \dfrac{IRL\omega}{(R^2 + \omega^2 L^2)}e^{-Rt/L}$,

$y = \dfrac{IL\omega}{\sqrt{R^2 + \omega^2 L^2}}\sin\left(\omega t - \phi + \dfrac{\pi}{2}\right) - \dfrac{IRL\omega}{R^2 + \omega^2 L^2}e^{-Rt/L}$,

$\phi = \tan^{-1}\left(\dfrac{\omega L}{R}\right)$

14 $x = \dfrac{I}{1 - \omega^2 LC}\left[\sin\omega t - \omega\sqrt{LC}\sin\dfrac{t}{\sqrt{LC}}\right]$,

$y = \dfrac{I\omega LC}{1 - \omega^2 LC}\left[-\omega\sin\omega t + \sqrt{LC}\sin\dfrac{t}{\sqrt{LC}}\right]$

15 $x = \dfrac{a}{\omega^2}(\omega t - \sin\omega t),\ y = \dfrac{a}{\omega^2}(1 - \cos\omega t)$

Index